Lower Secondary
Mathematics
8

Ric Pimentel
Frankie Pimentel
Terry Wall

The Publishers would like to thank the following for permission to reproduce copyright material.

p.112 Max Roser (2021) – "Literacy rates by age-group". Published online at OurWorldInData.org. Retrieved from: https://ourworldindata.org/literacy [Online Resource]; **p.113** "Population by broad age group, Europe, 1950-2015." Published online at OurWorldInData.org. Retrieved from: https://ourworldindata.org/age-structure [Online Resource]; **p.114** Population by broad age group, Sub-Saharan Africa, 1950-2015". Published online at OurWorldInData.org. Retrieved from https://ourworldindata.org/age-structure [Online Resource].

Photo acknowledgements

p.vii © Andersphoto/stock.adobe.com; **p.1** © Carlos Huang/Shutterstock.com; **p.195** © RuslanKphoto/stock.adobe.com; **p.209** © Nikolai Sorokin/stock.adobe.com; **p.210** © Look Aod 27/stock.adobe.com; **p.228** © Orangeberry/stock.adobe.com; **p.246** © Fotolia

Acknowledgements

Cambridge International copyright material in this publication is reproduced under licence and remains the intellectual property of Cambridge Assessment International Education.

End of section review questions (and sample answers) have been written by the authors. In assessment, the way marks are awarded may be different.

Third-party websites and resources referred to in this publication have not been endorsed by Cambridge Assessment International Education.

Every effort has been made to trace all copyright holders, but if any have been inadvertently overlooked, the Publishers will be pleased to make the necessary arrangements at the first opportunity.

Although every effort has been made to ensure that website addresses are correct at time of going to press, Hodder Education cannot be held responsible for the content of any website mentioned in this book. It is sometimes possible to find a relocated web page by typing in the address of the home page for a website in the URL window of your browser.

Hachette UK's policy is to use papers that are natural, renewable and recyclable products and made from wood grown in well-managed forests and other controlled sources. The logging and manufacturing processes are expected to conform to the environmental regulations of the country of origin.

Orders: please contact Hachette UK Distribution, Hely Hutchinson Centre, Milton Road, Didcot, Oxfordshire, OX11 7HH. Telephone: +44 (0)1235 827827. Email education@hachette.co.uk Lines are open from 9 a.m. to 5 p.m., Monday to Friday. You can also order through our website: www.hoddereducation.com

ISBN: 978 1 398 30199 3

© Ric Pimentel, Frankie Pimentel and Terry Wall 2021

First published in 2004
Second edition published in 2011
This edition published in 2021 by
Hodder Education,
An Hachette UK Company
Carmelite House
50 Victoria Embankment
London EC4Y 0DZ

www.hoddereducation.com

The authorised representative in the EEA is Hachette Ireland, 8 Castlecourt Centre, Dublin 15, D15 XTP3, Ireland (email: info@hbgi.ie)

Impression number   10 9 8 7

Year   2025

All rights reserved. Apart from any use permitted under UK copyright law, no part of this publication may be reproduced or transmitted in any form or by any means, electronic or mechanical, including photocopying and recording, or held within any information storage and retrieval system, without permission in writing from the publisher or under licence from the Copyright Licensing Agency Limited. Further details of such licences (for reprographic reproduction) may be obtained from the Copyright Licensing Agency Limited, www.cla.co.uk

Cover photo © byjeng - stock.adobe.com

Illustrations by Integra Software Services Pvt. Ltd, Pondicherry, India

Typeset in Palatino LT Std Light 11/13 by Integra Software Services Pvt. Ltd, Pondicherry, India

Printed in Great Britain by Bell and Bain Ltd, Glasgow

A catalogue record for this title is available from the British Library.

Contents

The units in this book have been arranged to match the Cambridge Lower Secondary Mathematics curriculum framework. Each unit is colour coded according to the area of the syllabus it covers:

■ Number

■ Geometry & Measure

■ Statistics & Probability

■ Algebra

How to use this book v

Introduction vii

Section 1

Unit 1	Multiplication and division	2
Unit 2	Hierarchy of quadrilaterals	15
Unit 3	Data collection and sampling methods	19
Unit 4	Parallelograms, trapezia and circles	25
Unit 5	Order of operations	36
Unit 6	Expressions, formulae and equations	41
Unit 7	Recording, organising and representing data	48
Unit 8	Properties of three-dimensional shapes	73
Unit 9	Factors and multiples	83
Unit 10	Complementary events	90
Section 1 Review		93

Contents

Section 2

Unit 11 Decimals and place value	96
Unit 12 Comparing and interpreting data	105
Unit 13 Transformation of 2D shapes	115
Unit 14 Fractions and decimals	129
Unit 15 Manipulating algebraic expressions	141
Unit 16 Combined events	148
Unit 17 Constructions, lines and angles	153
Unit 18 Algebraic expressions and formulae	175
Unit 19 Probability experiments	181
Unit 20 Equations and inequalities	184
Section 2 Review	192

Section 3

Unit 21 Describing sequences	196
Unit 22 Percentage increases and decreases	204
Unit 23 2D representations of 3D shapes	209
Unit 24 Functions	214
Unit 25 Geometry and translations	221
Unit 26 Squares, square roots, cubes and cube roots	229
Unit 27 Graphs and equations of straight lines	234
Unit 28 Distances and bearings	244
Unit 29 Ratio	249
Unit 30 Reading and interpreting graphs	256
Section 3 Review	261

Glossary	**264**
Index	**269**

How to use this book

To make your study of Cambridge Checkpoint Lower Secondary Mathematics as rewarding as possible, look out for the following features when you are using the book:

History of mathematics

These sections give some historical background to the material in the section.

These aims show you what you will be covering in the unit.

LET'S TALK
Talk with a partner or a small group to decide your answer when you see this box.

Worked example

These show you how you would approach answering a question.

KEY INFORMATION
These give you hints or pointers to solving a problem or understanding a concept.

These highlight ideas and things to think about

This book contains lots of activities to help you learn. The questions are divided into levels by difficulty. Green are the introductory questions, amber are more challenging and red are questions to really challenge yourself. Some of the questions will also have symbols beside them to help you answer the questions.

How to use this book

Exercise 15.2

1. Work out the answer to the following calculations. Show your working clearly and simplify your answers where possible.
   a. $\frac{2}{5} + \frac{1}{6}$
   b. $\frac{7}{12} + \frac{1}{5}$
   c. $\frac{9}{14} - \frac{2}{7}$
   d. $\frac{3}{13} - \frac{3}{26}$
   e. $\frac{1}{8} + \frac{5}{16} - \frac{5}{24}$
   f. $\frac{13}{18} - \frac{8}{9} + \frac{1}{6}$

2. Sadiq spends $\frac{1}{5}$ of his earnings on his mortgage. He saves $\frac{2}{7}$ of his earnings. What fraction of his earnings is left?

3. The numerators of two fractions are hidden as shown.

   $$\frac{\square}{8} + \frac{\square}{5} = \frac{23}{40}$$

   The sum of the two fractions is $\frac{23}{40}$. Calculate the value of both numerators.

Look out for these symbols:

⭐ This green star icon shows the thinking and working mathematically (TWM) questions. This is an important approach to mathematical thinking and learning that has been incorporated throughout this book.

Questions involving TWM differ from the more straightforward traditional question and answer style of mathematical learning. Their aim is to encourage you to think more deeply about the problem involved, make connections between different areas of mathematics and articulate your thinking.

🖩 This indicates where you will see how to use a calculator to solve a problem.

❌ These questions should be answered without a calculator.

🔊 This tells you that content is available as audio. All audio is available to download for free from www.hoddereducation.com/cambridgeextras

🔵 There is a link to digital content at the end of each unit if you are using the Boost eBook.

Introduction

This is the second of a series of three books which follows the Cambridge Lower Secondary Mathematics curriculum framework. It has been written by three experienced teachers who have lived or worked in schools or with teachers in many countries including England, Spain, Germany, France, Turkey, South Africa, Malaysia and the USA.

Students and teachers studying this course come from a variety of cultures and speak many different languages as well as English. Sometimes cultural and language differences make understanding difficult.

However mathematics is almost a universal language. $28 + 37 = 65$ will be understood in many countries where English is not the first language. A mathematics book written in Japanese will include algebra equations with x and y.

We should also all be very aware that much of the mathematics you will learn in this series of three books was first discovered, and built upon, by mathematicians from China, India, Arabia, Greece and Western European countries.

The Fibonacci sequence of numbers is 0,1,1,2,3,5,8,13,21….This sequence is found many times in nature.

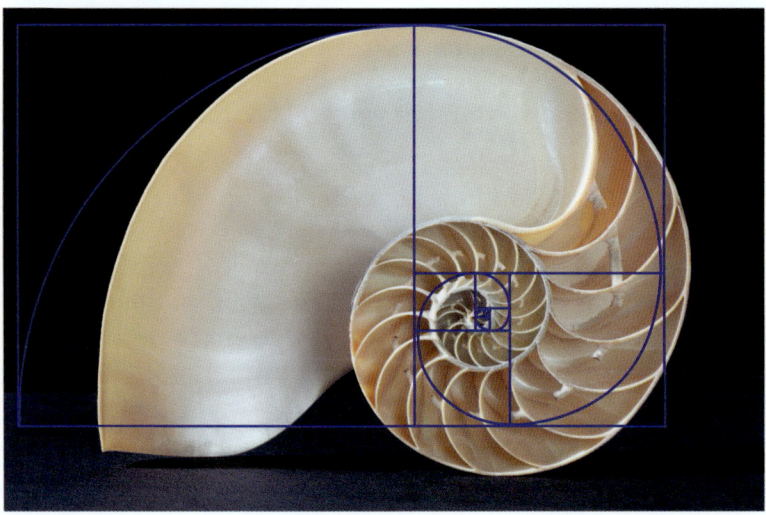

Introduction

Most early mathematics was simply game play and problem solving. It was not until later that this mathematics was applied to building, engineering and sciences of all kinds. Mathematicians study mathematics because they enjoy it as fun in itself.

We hope that you will enjoy the mathematics you learn in this series of books. Sometimes the ideas will not be easy to understand at first – that should be part of the fun! Ask for help if you need it, but try hard first.

Try to learn by Thinking and Working Mathematically (TWM). This is an important feature of the book. Thinking and Working Mathematically can be divided into the following characteristics:

Thinking and working characteristic	Definition
Specialising	Choosing an example and checking if it satisfies or does not satisfy specific mathematical criteria.
Generalising	Recognising an underlying pattern by identifying many examples that satisfy the same mathematical criteria.
Conjecturing	Forming mathematical questions or ideas.
Convincing	Presenting evidence to justify or challenge a mathematical idea or solution.
Characterising	Identifying and describing the mathematical properties of an object.
Classifying	Organising objects into groups according to their mathematical properties.
Critiquing	Comparing and evaluating mathematical ideas, representations or solutions to identify advantages and disadvantages.
Improving	Refining mathematical ideas or representations to develop a more elegant approach or solution

Where you see this icon ⭐ it shows you that you will be thinking and working mathematically.

Writing down your thoughts and workings helps to develop your mathematical fluency. By thinking carefully about how you explain your ideas you may, while justifying an answer, be able to make wider generalisations. Discussing different methods with other students will also help you compare and evaluate your mathematical ideas. This will lead to you understand why some methods are more effective than others in given situations. Throughout you should always be forming further mathematical questions and presenting other ideas for thought.

Many students start off by thinking that mathematics is just about answers. Although answers are often important, posing questions is just as important. What is certainly the case is that the more you question and understand, the more you will enjoy mathematics.

Ric Pimentel, Frankie Pimentel and Terry Wall, 2021

SECTION 1

History of mathematics – the Chinese (from 200BCE)

'For a civilization to endure and prosper, it must give its citizens order and fairness.'

Chang Tshang

Chang Tshang (ca 200–142BCE) was one of many Chinese mathematicians. They were the first to discover various algebraic and geometric ideas. The textbook 'Nine Chapters of the Mathematical Art' (known in Chinese as Jiu Zhang Suan Shu) has special importance. It was written during the early Han Dynasty (about 165BCE) by Chang Tshang.

Chang's book gives methods of arithmetic (including square and cube roots) and algebra (including a solution of equations). It uses the decimal system with zero and negative numbers. It has geometric proofs using triangles and circles.

Chang was concerned with the ordinary lives of the people, so his mathematical ideas were applied to science, building, agriculture and economics. He wrote 'For a civilization to endure and prosper, it must give its citizens order and fairness'. Three chapters were concerned with ratio and proportion, so that 'rice and other cereals can be planted in the correct proportion to our needs, and the ratio of taxes could be paid fairly'.

'Nine Chapters' was the main Chinese mathematical text for centuries and had great influence. Some of the teachings made their way to India, and from there to the Middle East and Europe. The Indian teachers may have borrowed the decimal system itself from books like 'Nine Chapters'.

# Multiplication and division

- Estimate, multiply and divide integers, recognising generalisations.
- Use positive and zero indices, and the index laws for multiplication and division.

Being able to **generalise** and work out calculations is possibly the most important aspect of mathematics at school. You will be taught different methods, some which you will like and understand, others maybe not so much.

This unit will look at some methods for estimating multiplications and divisions. Being able to estimate answers without the use of a calculator is a useful skill to have as it means you can spot errors.

The multiplication grid below is partially complete. It shows the results of multiplying two whole numbers together.

LET'S TALK
Explain to another student why multiplying a number by zero gives an answer of zero.

×	3	2	1	0	–1	–2
3	9	6	3	0		
2	6	4	2	0		
1	3	2	1	0		
0	0	0	0	0		
–1						
–2						

The top row of the table shows decreasing numbers being multiplied by three, giving the answers 9, 6, 3, 0. The answers clearly follow a pattern. The pattern can therefore be continued, and we can see what happens when we multiply a positive number by a negative number.

The same applies to the first four rows and columns:

×	3	2	1	0	–1	–2
3	9	6	3	0	–3	–6
2	6	4	2	0	–2	–4
1	3	2	1	0	–1	–2
0	0	0	0	0	0	0
–1	–3	–2	–1	0		
–2	–6	–4	–2	0		

1 Multiplication and division

If the patterns are continued further, then we can see what happens when two negative numbers multiply each other:

×	3	2	1	0	−1	−2
3	9	6	3	0	−3	−6
2	6	4	2	0	−2	−4
1	3	2	1	0	−1	−2
0	0	0	0	0	0	0
−1	−3	−2	−1	0	1	2
−2	−6	−4	−2	0	2	4

> The last of the three statements in particular, a lot of people get wrong and make mistakes with.

From the grids we can conclude that:
- When two positive numbers are multiplied together, we get a *positive* answer.
- When a positive number and a negative number are multiplied together, we get a *negative* answer.
- When two negative numbers are multiplied together, we get a *positive* answer.

Worked example

a Multiply
  i) 50×7
     We know that $5 \times 7 = 35$
     50 is ten times bigger than 5
     So 50×7 must be ten times bigger than 5×7
     $50 \times 7 = 35 \times 10 = 350$
  ii) 51×7
     51×7 can be visualised as shown in the rectangle below:

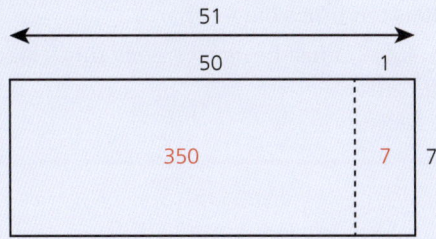

     We have already found that $50 \times 7 = 350$
     $51 \times 7 = (50 \times 7) + (1 \times 7) = 350 + 7 = 357$

SECTION 1

iii) -51×7

-51×7 when multiplied must give a negative answer because we are multiplying a positive number and a negative number together. The answer will have the same size as 51×7 but *a different* sign.

$-51 \times 7 = -357$

iv) $-51 \times (-7)$

$-51 \times (-7)$ must give a positive answer because we are multiplying two negative numbers together.

$-51 \times (-7) = 357$

> 357 and −357 are the same size of number but opposite signs. One is positive, the other negative.

b **Multiply**

i) 70×80

We know that $7 \times 8 = 56$.

70 is ten times bigger than 7 and 80 is ten times bigger than 8.

So 70×80, which can be written as $7 \times 10 \times 8 \times 10$, must be one hundred times bigger than 7×8.

$70 \times 80 = 56 \times 100 = 5600$

ii) 72×80

Visualising 72×80 as a rectangle: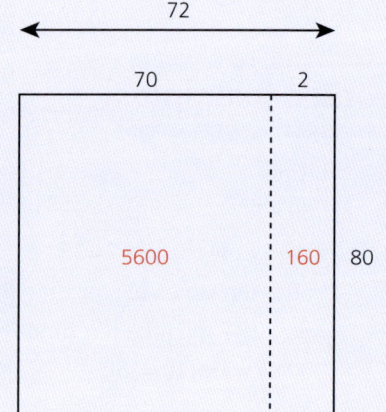

We already know that $70 \times 80 = 5600$.

$72 \times 80 = (70 \times 80) + (2 \times 80) = 5600 + 160 = 5760$

iii) $72 \times (-80)$

$72 \times (-80)$ will have the same size answer as 72×80 but a different sign.

Therefore, $72 \times (-80) = -5760$.

c **Multiply**

i) 600×40

We know that $6 \times 4 = 24$.

600 is one hundred times bigger than 6 and 40 is ten times bigger than 4.

So, 600×40, which can be written as $6 \times 100 \times 4 \times 10$, must be *one thousand* times bigger than 6×4.

$600 \times 40 = 24 \times 1000 = 24\,000$

ii) 602×43

To calculate 602×43 look at the rectangle below:

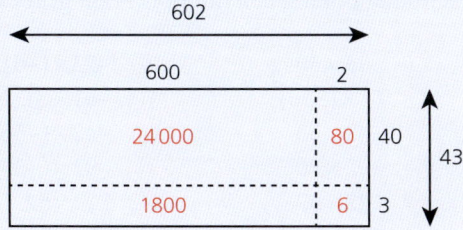

We have found that $600 \times 40 = 24\,000$.

$602 \times 43 = (600 \times 40) + (2 \times 40) + (600 \times 3) + (2 \times 3)$

We have found that $600 \times 40 = 24\,000$.

$602 \times 43 = (600 \times 40) + (2 \times 40) + (600 \times 3) + (2 \times 3)$
$= 24\,000 + 80 + 1800 + 6 = 25\,886$

iii) $(-301) \times 43$

To calculate $(-301) \times 43$ does not need starting from scratch.

Multiplying a positive and a negative number together will produce a negative answer. 301 is half of 602 therefore the size of the answer will be halved too.

$(-301) \times 43 = -\frac{25\,886}{2} = -12\,943$

iv) $(-301) \times (-86)$

Once again, to calculate $(-301) \times (-86)$ study how it compares with the earlier calculations.

Compared with 602×43, we can see that $(-301) \times (-86)$ has one number that is halved and the other which has been doubled. As doubling and halving are opposite operations their effects cancel out. As both numbers are negative, the result of multiplying them together is positive.

Therefore, $(-301) \times (-86) = 602 \times 43 = 25\,886$.

LET'S TALK
What other calculations will give the same answer as 602×43?

SECTION 1

Exercise 1.1

Multiply the following pairs of numbers mentally, without looking at a multiplication grid or using a calculator. Write down your answers.

1. a) 20 × 5 b) 23 × 5 c) −23 × 50 d) −23 × (−51)
2. a) 60 × 3 b) 61 × 3 c) 61 × (−30) d) −61 × 31
3. a) 40 × 5 b) 42 × 5 c) −42 × 53 d) −42 × (−53)
4. a) −9 × 30 b) 90 × 30 c) −92 × 30 d) −92 × (−33)
5. a) 20 × 5 b) −22 × (−5) c) −22 × 52 d) 220 × 52
6. a) 60 × (−9) b) −61 × 9 c) −61 × (−95) d) 61 × 95

7. The following cards show 10 calculations.

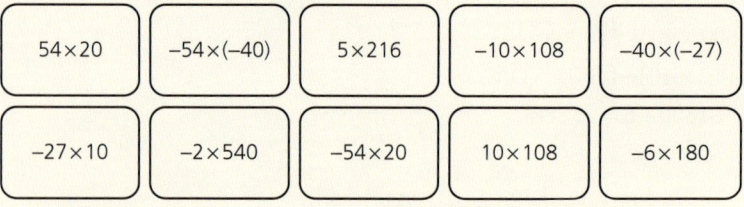

a **Classify** the cards into groups giving the same answer.
b i) State which two calculations do not belong to either group.
  ii) Work out the answers to these two calculations.

8. Given that 78 × 24 = 1872 state which of the following are correct and which are incorrect. Justify your answers in each case.
a 78 × 12 = 936
b −78 × (−24) = −1872
c 39 × 48 = 936
d −156 × 12 = −936

> Note that $\frac{240}{5} = 48$ can also be written as $240 \div 5 = 48$. Similarly, $\frac{240}{48} = 5$ can be written as $240 \div 48 = 5$.

Calculation patterns can also be used to deduce the answers to divisions.

For example, as 48 × 5 = 240 this implies that $\frac{240}{5} = 48$.

It also implies that $\frac{240}{48} = 5$.

This helps us work out what happens when negative numbers are used.

We know that −48 × 5 = −240, because multiplying a negative number by a positive number gives a negative answer.

Rearranging, this can be written as a division as $\frac{-240}{5} = -48$ or $\frac{-240}{-48} = 5$.

Therefore we can see, as with multiplication, that:
- dividing a negative number by a positive number (or vice versa) produces a negative answer
- dividing a negative number by another negative number produces a positive answer.

1 Multiplication and division

Worked examples

1. Given that $55 \times 18 = 990$, identify which of the following calculations must be wrong. Justify your answers.

   a $55 \times (-18) = -990$

   b $\frac{990}{-18} = 55$

   c $\frac{-990}{-55} = -18$

   d $-55 = \frac{-990}{18}$

   e $\frac{990}{-55} = -18$

   f $\frac{55}{990} = 18$

   (b) is incorrect as dividing a positive number by a negative number must give a negative answer.

   (c) is incorrect as dividing a negative number by another negative number must produce a positive answer.

   (f) is incorrect because the numerator is smaller than the denominator, therefore the answer cannot be bigger than 1.

2. Given that $\frac{360}{4} = 90$ deduce and give **convincing** answers to the following.

   a $\frac{360}{8}$

   $\frac{360}{8} = 45$ As the denominator has been doubled, the answer must be halved.

   This can be explained by looking at fractions. $\frac{360}{8} = \frac{1}{2} \times \frac{360}{4} = \frac{1}{2} \times 90 = 45$

   b $\frac{-360}{4}$

   $\frac{-360}{4} = -90$ The size of the numbers are unchanged from the original calculation, but a negative number is now being divided by a positive number, so the answer must be negative.

   c $\frac{720}{8}$

   $\frac{720}{8} = 90$ As both numerator and denominator have doubled, the answer will remain the same.

   This can be explained using fractions. $\frac{720}{8} = \frac{2}{2} \times \frac{360}{4} = 1 \times \frac{360}{4} = 1 \times 90 = 90$

   d $\frac{-720}{-2}$

   $\frac{-720}{-2} = 360$ As a negative number is being divided by a negative number, the answer must be positive. When compared with the original calculation, the size of the numerator has been doubled and the size of the denominator halved, therefore the answer will be four times bigger.

   This can be explained as follows: $\frac{-720}{-2} = \frac{720}{2} = 720 \div 2 = 360$.

3. a Show that an estimate for $6086 \div 29$ is 200.

   $\frac{6086}{29}$ can be approximated to $\frac{6000}{30} = 200$.

   b Estimate the answers to the following

   i) $\frac{-6086}{-29}$

   As both numbers are negative the answer will be positive.

   Therefore, $\frac{-6086}{-29} = \frac{6086}{29} \approx 200$.

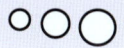

The symbol for 'approximately equals to' is ≈.

Therefore $\frac{6086}{29} \approx 200$.

SECTION 1

ii) $\frac{-6086}{61}$

This can be estimated in two ways.

By writing each number to make an easier calculation we can write:

$\frac{-6086}{61} \approx \frac{-6000}{60} = -100$

Or we can compare $\frac{-6086}{61}$ to the original calculation of $\frac{6086}{29}$.

It can be seen that the denominator has roughly doubled and therefore the answer will roughly halve. Also, as a negative number is being divided by a positive number, the answer will be negative.

Therefore, $\frac{-6086}{61} \approx \frac{1}{2} \times \frac{-6000}{30} = \frac{1}{2} \times (-200) = -100$.

Exercise 1.2

1 Work out the following divisions

   a i) $42 \div 6$ iii) $-42 \div 7$

        ii) $42 \div 7$ iv) $-42 \div (-6)$

   b i) $81 \div 9$ iii) $-81 \div 9$

        ii) $81 \div (-9)$ iv) $(-81) \div (-9)$

   c i) $\frac{225}{25}$ iii) $\frac{-225}{-25}$

        ii) $225 \div 50$ iv) $450 \div (-25)$

   d i) $\frac{64}{8}$ iii) $\frac{-256}{-8}$

        ii) $\frac{128}{-8}$ iv) $\frac{256}{32}$

2 **a** Find the value of each of the letters below, so that the calculation gives the same answer as $\frac{36}{18}$.

      i) $\frac{-36}{a}$ iii) $\frac{-72}{-c}$ v) $\frac{e}{-72}$

      ii) $\frac{b}{9}$ iv) $\frac{-6}{d}$ vi) $\frac{-720}{-f}$

   b Calculate the value of each letter if the answer is to be double $\frac{36}{18}$.

3 The calculation $\frac{324}{9} = 36$ is shown in the centre of the spider diagram below.

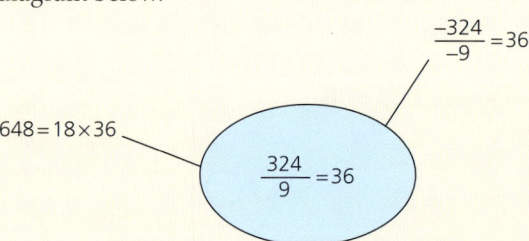

> A spider diagram has a central body and off it the 'legs'.
>
> In this case the calculations on each leg are related to the calculation in the centre.

Copy the spider diagram and write as many calculations as possible that can be deduced from the one in the middle. Two examples have been given.

You need to use your **specialising** skills to check each of your calculations can be found from the one in the centre.

1 Multiplication and division

4 Estimate the value of the missing quantity in each of the following:

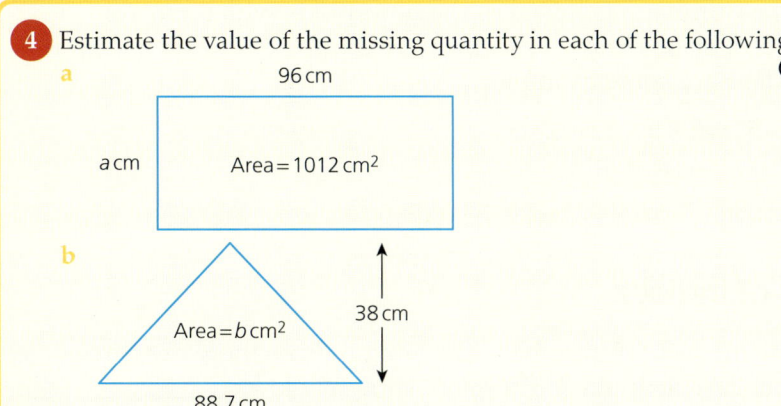

When estimating an answer you should not work out the exact answer.

 **5** A roll of tape measures 1893 m.
Eighteen lengths of 87 m each are cut from the roll.
  a Using numbers, write down the calculation needed to work out the amount of tape left on the roll.
  b Using your answer to part (a), estimate the amount of tape left on the roll
  c Estimate how many more of the same lengths could be cut from the roll. Justify your answer.
  d Check your answers to parts (b) and (c) above using a calculator and comment on the accuracy of your answers.

Indices

You have encountered some indices before when studying square and cube numbers in Stage 7.

Al-Karaji was one of the greatest Arab mathematicians. He lived in the 11th century. He wrote many books on algebra and developed a theory of indices and a method of finding square roots.

In the expression $ax^4 + bx^3 + cx^2 + dx$, the small numbers 4, 3 and 2 are called indices. Indices is the plural of index. So

x^4 has index 4

x^3 has index 3

x^2 has index 2

Although x does not appear to have an index, in fact it has index 1 but this is not usually written. So $x = x^1$ has index 1.

The index is the power to which a number is raised.

9

SECTION 1

In 5^3, the number 5 is raised to the power of 3, which means $5 \times 5 \times 5$. The 3 is known as the index and the 5 is known as the **base**. Here are some examples.

$$5^3 = 5 \times 5 \times 5 = 125$$
$$7^4 = 7 \times 7 \times 7 \times 7 = 2401$$
$$3^1 = 3$$

> **KEY INFORMATION**
> Your calculator will have an index button. It may look like x^y or simply ^.
>
> Make sure you know which one it is and how it works.

Laws of indices

When working with numbers involving indices there are three basic laws that can be applied. These are shown below.

- $4^2 \times 4^4 = 4 \times 4 \times 4 \times 4 \times 4 \times 4$
  $= 4^6$ (i.e. 4^{2+4})

This can be **generalised** as:

$$a^m \times a^n = a^{m+n}$$

Notice that the base numbers must be the same for this rule to be true.

- $3^6 \div 3^2 = \dfrac{3 \times 3 \times 3 \times 3 \times \cancel{3} \times \cancel{3}}{\cancel{3} \times \cancel{3}}$
  $= 3^4$ (i.e. 3^{6-2})

This can be **generalised** as:

$$a^m \div a^n = a^{m-n}$$

Again, the base numbers must be the same for this rule to be true

- $\left(5^2\right)^3 = (5 \times 5) \times (5 \times 5) \times (5 \times 5)$
  $= 5^6$ (i.e. $5^{2 \times 3}$)

This can be **generalised** as:

$$\left(a^m\right)^n = a^{mn}$$

Worked example

a Simplify $4^3 \times 4^2$.
$4^3 \times 4^2 = 4^{(3+2)}$
$= 4^5$

b Evaluate $\left(4^2\right)^3$.
$\left(4^2\right)^3 = 4^{(2 \times 3)}$
$= 4^6$
$= 4096$

c Simplify $2 \times 2 \times 2 \times 5 \times 5$ using indices.
$2 \times 2 \times 2 \times 5 \times 5 = 2^3 \times 5^2$

1 Multiplication and division

Exercise 1.3

1 Write the following using indices.
  a $4 \times 4 \times 4$
  b $3 \times 3 \times 3 \times 3 \times 3$
  c $7 \times 7 \times 7 \times 7 \times 7 \times 7$
  d 6×6
  e $12 \times 12 \times 12$

2 Write out the following in full.
  a 7^4
  b 3^3
  c 9^4
  d 6^5
  e 11^2

3 Simplify the following using indices.
  a $2^3 \times 2^2$
  b $3^4 \times 3^5$
  c $4^2 \times 4^3 \times 4^4$
  d 5×5^2
  e $8^3 \times 8^2 \times 8$
  f $6^3 \div 6^2$
  g $8^5 \div 8^2$
  h $2^7 \div 2^6$
  i $10^5 \div 10^3$
  j $3^9 \div 3$

4 Simplify the following.
  a $(4^3)^2$
  b $(3^2)^3$
  c $(2^5)^2$
  d $(4^3)^4$
  e $(3^7)^2$
  f $(2^4)^4$
  g $(5^2)^2$
  h $(6^3)^2$
  i $(7^2)^4$
  j $(8^4)^4$

5 A student is trying to work out the following calculation using the laws of indices.
$3^4 \times 3^6 \times 3$
He says that this can be worked out as
$3^{4+6+0} = 3^{10}$.
  a Explain why his calculation is incorrect.
  b Write down the correct answer using indices.

6 A square has side lengths of 2 cm as shown. Four of these squares are arranged to form a larger square.

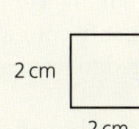
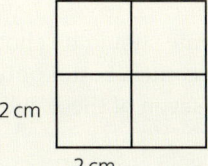

2 cm   2 cm
2 cm   2 cm

  a Explain why the area of the large square can be expressed as 2^4 cm^2.
  b How many smaller squares are arranged to form a larger square with an area of 2^6 cm^2? Give a **convincing** justification for your answer.

SECTION 1

7 Simplify the following.
- a $2^3 \times 2^4$
- b $3^5 \div 3^2$
- c $2^4 \times 2^2 \div 2^3$
- d $3^8 \times 3^2 \div 3^4$
- e $5^2 \times 5^3 \div 5$
- f $6^3 \times 6 \div 6^2$
- g $(3^4)^2 \div 3^3$
- h $(5^5)^2 \div 5^3$
- i $(6^2)^3 \div 6^3$
- j $(7^3)^4 \div 7^5$

8 Scientists are looking at the rate at which flies multiply. At the start of their data collection there are 128 flies. They notice that the population doubles every day.
After how many days will the fly population number 2^{16}? Justify your answer.

If the base numbers are not the same, only parts of the expression can be simplified. For example,

$5 \times 5 \times 5 \times 5 \times 5 \times 3 \times 3 = 5^5 \times 3^2$

Exercise 1.4

1 Simplify the following. Leave your answers in index form.
- a $3 \times 3 \times 2 \times 2$
- b $4 \times 4 \times 5 \times 5 \times 5$
- c $2 \times 2 \times 3 \times 3 \times 3$
- d $3 \times 1 \times 4 \times 5 \times 5$
- e $5 \times 5 \times 5 \times 5 \times 6 \times 6$
- f $7 \times 7 \times 4 \times 4 \times 4$
- g $2 \times 2 \times 3 \times 3 \times 5 \times 5$
- h $2 \times 3 \times 3 \times 5 \times 5 \times 5$

2 A student decides to take on a job to help a neighbour clear their garden. The neighbour offers two different pay rates.
Rate A: Get paid $15 for each hour
Rate B: Get paid $2 for one hour, $4 for two hours, $8 for three hours and so on, doubling the amount each hour that he works.
- a Write a formula for the total amount (P) he would receive for doing n hours of work using rate A.
- b Explain why the amount he would be paid using rate B can be calculated using the formula $P = 2^n$.
- c i) If he works for four hours, what pay rate should he choose? Justify your answer.
     ii) After how many hours does rate B become a better option than A? Justify your answer.

3 A cube has a side length of 3 units. Twenty-seven of these cubes are arranged to form a larger cube.
- a Explain, with the aid of a diagram, why the total surface area of the large cube can be given by the expression 6×3^4 units^2.
- b Write and expression for the volume of the large cube in the form 3^x where x is an integer.

The zero index

The **zero index** means that a number has been raised to the power of 0. Any number raised to the power of 0 is equal to 1. For example,

$$4^0 = 1 \qquad 10^0 = 1 \qquad a^0 = 1$$

This can be explained by applying the laws of indices.

$$a^m \div a^n = a^{m-n}$$

Therefore,

$$\frac{a^m}{a^m} = a^{m-m}$$
$$= a^0$$

But

$$\frac{a^m}{a^m} = 1$$

Therefore,

$$a^0 = 1$$

This can also be demonstrated using numbers:

$$5^3 = 5 \times 5 \times 5$$
$$5^2 = 5 \times 5$$
$$5^1 = 5$$
$$5^0 = 1$$

$\div 5$

Worked example

Using indices, find the value of each of these.

a $\quad 3^6 \div 3^4$

$\quad 3^6 \div 3^4 = 3^2 = 9$

b $\quad 4^5 \div 4^3$

$\quad 4^5 \div 4^3 = 4^2 = 16$

SECTION 1

Exercise 1.5

Using indices, find the value of each of these.

1.
   a $5^3 \div 5^2$
   b $3^8 \div 3^6$
   c $4^2 \times 4^3 \div 4^4$
   d $2^4 \times 2^2 \div 2^6$
   e $3^3 \times 3^4 \div 3^7$
   f $2^4 \div 2^2$
   g $3^4 \div 3^3$
   h $4^4 \div 4^4$
   i $5^2 \times 5^5 \div 5^6$
   j $(6^2)^3 \div (6^3)^2$

LET'S TALK
Without calculating the values can you explain whether $(5^4)^3$ is the same as $(5^3)^4$?

2. Five calculations involving indices are shown below.

   $(4^0)^2 \times (3^3)^0$ $\dfrac{5^0}{(6^2)^0 \times 1^4}$ $1^6 \times (x^2)^0$

   $\dfrac{10^1 \times (3^0)^2}{2 \times 5^0}$ $\dfrac{3^5 \div 9}{3^3}$

   a i) Which calculation produces a different value from the other four?
       ii) What value does the odd one out give?
   b What value do the four others give?

▶ Now you have completed Unit 1, you may like to try the Unit 1 online knowledge test if you are using the Boost eBook.

2 Hierarchy of quadrilaterals

- Identify and describe the hierarchy of quadrilaterals.

Hierarchy of quadrilaterals

You will have seen in Science that zoologists sometimes use a flow chart to categorise or **classify** animals into certain groups. This is useful because it identifies what characteristics make one animal group different from another, but also what similarities they have and what properties they share.

Below is a simple flow chart which can be used to **classify** the eight animals shown.

> **LET'S TALK**
> Which of the eight animals belong to each of the boxes A, B, C and D?
>
> What additional characteristic would help separate the animals in each box?

> **KEY INFORMATION**
> An invertebrate is an animal without a backbone. A vertebrate is an animal with a backbone.

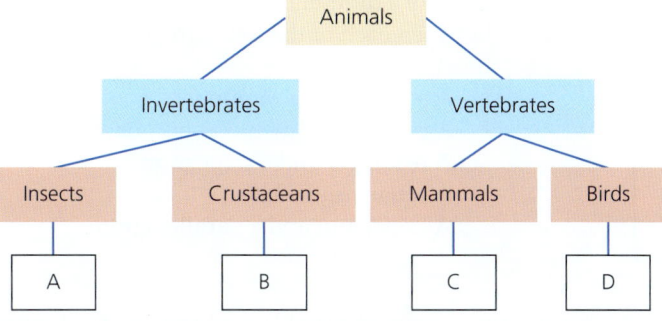

The flow chart shows that there is a **hierarchy**. The group 'mammals' belongs to the group 'vertebrate' which in itself belongs to the group of 'animals'. To be able to separate the different animals within the group 'mammals' needs further characteristics, e.g. carnivores and herbivores.

SECTION 1

In mathematics, shapes can be identified and **classified** according to their **characteristics** too. This unit looks at quadrilaterals in particular and the hierarchy within them.

The seven quadrilaterals being looked at are shown below:

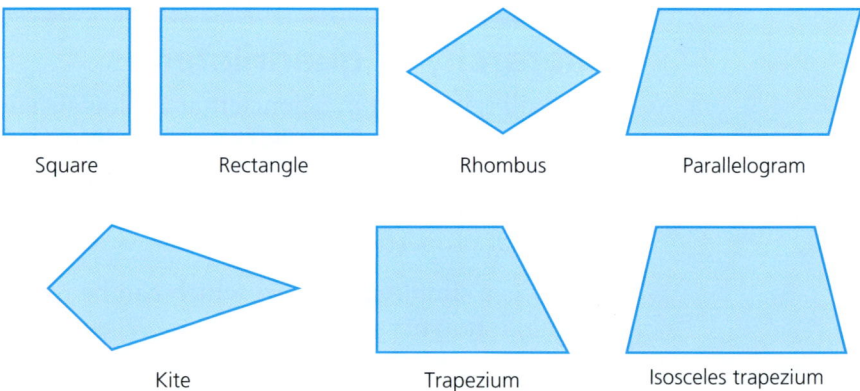

Square   Rectangle   Rhombus   Parallelogram

Kite   Trapezium   Isosceles trapezium

> **Adjacent** sides are next to each other.

Exercise 2.1

1.
   a. Which of the seven quadrilaterals have at least one pair of parallel sides?
   b. What additional description could divide the group further?

2.
   a. Which of the seven quadrilaterals have at least one pair of adjacent sides of the same length?
   b. What additional description could divide the group further?

3. Write a description that would
   i) identify a similarity
   ii) identify a difference
   between each of the following pairs of quadrilaterals:
   a. A square and a rhombus
   b. A parallelogram and a rhombus
   c. A rectangle and a trapezium
   d. A square and a kite
   e. A trapezium and a parallelogram

> **LET'S TALK**
> Can you find more than one similarity and difference between each pair of quadrilaterals in question 3?

> You need to use:
> - **characterising** skills to describe the properties of each shape
> - **classifying** skills to organise the shapes into groups.

From your answers you should realise that some properties, because they belong to more than one shape, are not enough on their own to help distinguish them from each other. Therefore further properties are sometimes necessary.

2 Hierarchy of quadrilaterals

This is a quadrilateral hierachy diagram:

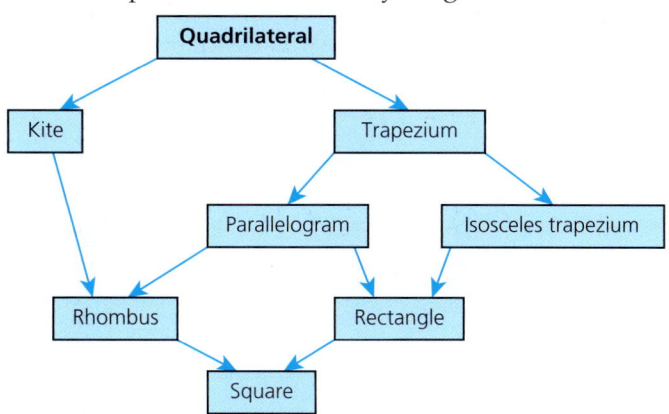

At the top is 'Quadrilateral', which means that all the shapes linked below it are special types of quadrilateral. This continues all the way to the bottom of the diagram. Therefore, those shapes which are below 'Parallelogram' are special types of parallelogram.

Exercise 2.2

1. Using the diagram above, decide which shape(s) are a **special** type of
   a. rhombus
   b. kite
   c. parallelogram
   d. isosceles trapezium.

2. Natalia says that the hierarchy of quadrilaterals diagram must be wrong because it is saying that a square is a type of parallelogram and yet they look completely different.
   How would you **convince** Natalia that she is wrong?

3. Explain, using diagrams of quadrilaterals to help, the meaning of the following mathematical words
   a. adjacent sides
   b. parallel sides
   c. opposite angles
   d. diagonals
   e. equal sides.

4. Identify which quadrilaterals fit the following **classifications**:
   a. All four sides are congruent.
   b. The diagonals intersect at right angles.
   c. At least one pair of parallel sides.
   d. Opposite sides parallel.
   e. Have four right angles.

17

SECTION 1

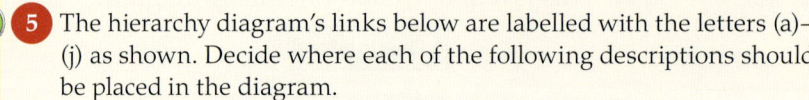

5   The hierarchy diagram's links below are labelled with the letters (a)–(j) as shown. Decide where each of the following descriptions should be placed in the diagram.

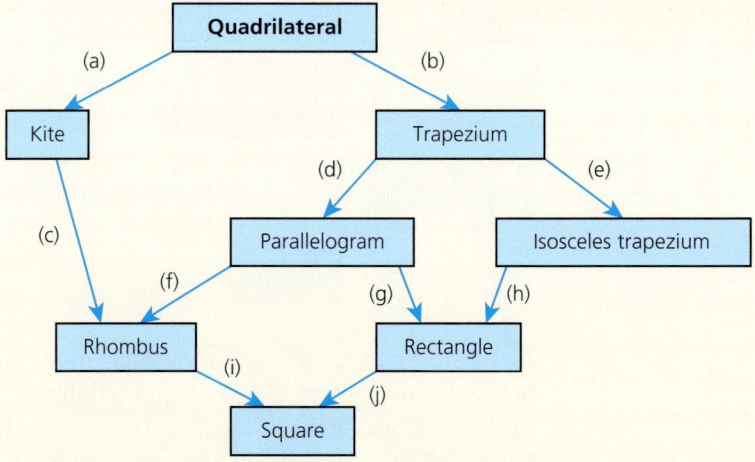

> Remember that each statement further down the diagram must add an extra description to narrow the choice. It must not repeat a description that has already been stated further up.

1   Four right angles and congruent diagonals.
2   Two pairs of congruent adjacent sides and their diagonals intersect at right angles.
3   Two pairs of congruent sides and four right angles.
4   At least one pair of parallel sides.
5   All sides are congruent and opposite angles equal.
6   Diagonals cross at right angles.
7   Opposite sides parallel and opposite angles equal.
8   Four right angles.
9   One pair of congruent sides and two pairs of congruent adjacent angles.
10 All sides are congruent.

LET'S TALK

Are there other descriptions which could be used to categorise the seven quadrilaterals according to this hierarchy?

Now you have completed Unit 2, you may like to try the Unit 2 online knowledge test if you are using the Boost eBook.

3 Data collection and sampling methods

- Select, trial and justify data collection and sampling methods to investigate predictions for a set of related statistical questions, considering what data to collect (categorical, discrete and continuous data).
- Understand the advantages and disadvantages of different sampling methods.

> *Remember, quantitative data are data that can be measured. Discrete data takes specific values, while continuous data can take any value, usually within a range.*

Sampling methods

In Stage 7, you studied the different types of data which could be collected. These included **quantitative data**, which itself can be divided into **discrete data** and **continuous data**.

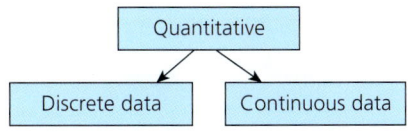

> *Remember, categorical data are data which can be put into groups or categories.*

Another type of data considered was **categorical data**.

> **LET'S TALK**
> Can you name at least two examples of data for each of the three types mentioned?

KEY INFORMATION
The word population in data collection does not mean everybody in the country. Depending on the context, the whole population could mean all the children of a particular age group, or just all the children in a particular class.

Collecting accurate data is an important area of mathematics and is used a lot in the modern world. Data for medical research can tell doctors how effective a cancer treatment is. Data for the masses of children at different stages of growth can identify children who might be under- or overweight. Data for CO_2 emissions can help scientists model the effects of climate change in the future.

When data are collected, it is unlikely that a whole **population** will be included. Usually a **sample** of the population is used. The important thing is that the sample should be **representative** of the whole population. A representative sample should give similar results to the whole population.

To do this several different types of sampling have been developed. In this section, you will trial some of these methods and decide whether they are representative. You will need the resource sheet containing details for 30 students from your teacher.

SECTION 1

LET'S TALK
Although the students are randomly placed, hair colour and eye colours do not appear to be random. Discuss why this might be the case.

They include the following data as shown in the example:
- First name
- Gender
- Hair colour
- Eye colour
- Height

Isabel	
Gender:	Girl
Hair:	Brown
Eyes:	Blue
Height:	152 cm

The 30 students have been arranged randomly on the sheet.

In order to investigate different sampling methods, we will compare their results to that of the whole population of 30 students.

Exercise 3.1

1 For the whole population of 30 students, copy and complete the following frequency tables.

a

Hair colour	Girls	Boys	Total
Black			
Brown			
Blonde			
Ginger			
Total			

b

		Eye colour			
		Brown	Blue	Green	Total
Hair colour	Black				
	Brown				
	Blonde				
	Ginger				
	Total				

2 Calculate the mean height of the girls and boys.

3 The data types mentioned are either discrete, continuous or categorical. Which type of data is:
  a gender
  b hair colour
  c eye colour
  d height?
Justify your answer.

3 Data collection and sampling methods

> **KEY INFORMATION**
> Calculators can generate numbers randomly usually using a Rand button.
> Using a Rand button usually generates a decimal between 0 and 1.
>
> Other scientific calculators have a RandInt button. Typing RandInt(1,30) will generate integers between 1 and 30.

First decide on a **sample size**. This is the number from the population that will be chosen as a sample.

> **LET'S TALK**
> In this case, with a population of 30 what might be a suitable sample size and why?

The simplest way of sampling a population is to carry out a random sample.

This can be done by putting all the student's names in a container and then picking them out one by one without looking.

Alternatively, you can number each student 1–30 and then get the random number generator on your calculator to choose the numbers.

Exercise 3.2

1. Choose a random sample of 10 students from the population. Explain how you selected your 10 students.
2. a Repeat questions 1 and 2 from Exercise 3.1 for your sample.
   b Comment on how the results of your sample
      i) differ from the population results
      ii) are similar to the population results.

> **LET'S TALK**
> Do you think any sample of 10 students would produce similar results?

Another type of sampling is to use a system for selecting the sample. To do this could involve arranging all the students in alphabetical order and then selecting every third student. Choosing every third student would mean getting a sample size of 10 students as before.

Exercise 3.3

1. a i) Choose a sample of 10 students from the population using a system.
      ii) Describe the system you have used.
   b Which 10 students are in your sample?
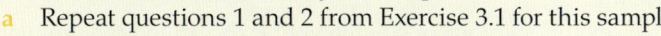2. a Repeat questions 1 and 2 from Exercise 3.1 for this sample.
   b Comment on how the results of your sample:
      i) differ from the population results
      ii) are similar to the population results.
   c Has using a system for sampling **improved** your results by producing a more representative sample of the whole population than the simple random sample? Give a **convincing** reason for your answer.

SECTION 1

> **LET'S TALK**
> If a sample contains more boys than girls how might this affect the results compared to the whole population?

You will have noticed that in the population of 30 students, 18 were girls and only 12 were boys. You will have seen in question 2 of Exercise 3.1 that the mean height of boys is greater than the mean height of girls. If a sample of these students happens to have more boys than girls then it is likely to affect the results.

One way to take into account the different number of boys and girls in your sample is to sample students in the same **ratio** as the population.

Exercise 3.4

1. **a** What is the ratio of girls to boys in the population? Give your answer in its simplest form.
   b **i)** If for a sample of 10, 5 boys and 5 girls were chosen, explain why this may produce results that are not representative of the population.
   ii) How could the sampling be **improved**?
   c If a sample size of 10 is wanted, approximately how many girls and boys should be chosen?
   d How do the number of girls and boys chosen using this ratio differ from the numbers chosen in your random sample in Exercise 3.2 and in your system method in Exercise 3.3?

2. **a** By using a random sample of 10 students chosen in the same gender ratio as the population, repeat questions 1 and 2 from Exercise 3.1.
   b Comment on how the results of your sample:
   i) differ from the population results
   ii) are similar to the population results.
   c Has using this method of sampling **improved** your results by producing a more representative sample of the whole population than the two other sampling methods? Justify your answers.

3. Does sample size affect the accuracy of your findings compared with the population results? Justify your answer giving full explanations.

4. Below are a number of data collection scenarios. How should the population be sampled in each case? **Critique** your chosen sampling method, making clear any assumptions you make.
   a A political party wants to know people's opinions on a new local policy idea. They decide to stand in a town centres and ask people as they walk by. They cannot ask everyone.
   b A television streaming service wants to ask its subscribers about the sorts of programmes they would like to see in the future. They have 5 million subscribers but they cannot ask all of them.
   c A football club wants to attract more female supporters to the club as currently only 10% of their supporters are female. They decide to ask the existing supporters for their ideas, but they cannot afford to ask them all.
   d A local newspaper wants to survey the opinions of cinema goers who attend the premier of a new animated film. They decide to survey the viewers at one cinema. They know in advance that of the 500 people who attended 350 were children.

> **LET'S TALK**
> In the examples above, can any of the other sampling methods be used? Are any definitely not appropriate?

3 Data collection and sampling methods

> **KEY INFORMATION**
> A biased question means that the person answering the question is being influenced by the way the question is written.

Questionnaires and interviews

You studied in Stage 7 how questionnaires are often used for data collection. This is because they are easy to produce and lots of data can be generated quickly and cheaply. However, using questionnaires poses its own problems.

The three main ones are:
- When designing them, you must make sure questions are not biased.
- People must interpret the question the way it was intended.
- The answer choices must be clear.

Interviews are sometimes used as way of overcoming some of the issues of using a questionnaire. If a question or answer is unclear, the interviewer can repeat it in a different way or ask the interviewee to clarify their answer.

Worked example

A clothing manufacturer wants to produce a questionnaire to help decide what to design next.

The first three questions of the questionnaire are:
1. Most normal people LOVE fashion. Are you interested in fashion?
   Yes ☐   No ☐
2. What type of clothing do you prefer?
   Trousers ☐   Skirts ☐   Jumpers ☐
3. How old are you?
   Young ☐   Old ☐

a Describe the reasons why these questions may not be very good.

Question 1 is a biased question. Saying that 'normal' people love fashion implies that you are not normal if you are not interested in it. People may therefore be reluctant to choose 'no'.

Question 2 has many flaws, including the following:
i) There are insufficient options and one option should say 'other'.
ii) Are men also going to be asked? Men are less likely to choose skirts as an option, so bias can be introduced with only these three options.
iii) When are people going to be asked? Jumpers are more likely to be chosen in Winter than in Summer.

Question 3 can be interpreted in different ways by different people. What is old and what is young?

> If a questionnaire is not trialled before it is used with hundreds of people, then the results may be meaningless if later it is found that there are problems with the way questions are written.

SECTION 1

A 40-year-old may consider himself to be young, but the person designing the questionnaire may have thought that anyone over 30 is old.

When designing a questionnaire, it is therefore important to trial it to see if there are any problems with the questions. Once a questionnaire has been trialled, **improvements** can be made to correct any unexpected problems that have occurred.

b Give one advantage and one disadvantage in collecting this information using an interview rather than a questionnaire.

An advantage would be that the flaws identified above would soon become obvious when speaking to people, therefore the interviewer could adapt their questioning straight away.

A disadvantage would be that the responses would take longer to analyse after the interview has finished.

Exercise 3.5

The following question is a mini-project and you should do this in pairs or small groups over an extended period of time to fully engage with the mathematics involved.

1 Decide on an important issue that you think affects the whole school. Your choice can be as a result of having interviewed people to see what they consider important.
Design a questionnaire with approximately 10 questions. Then carry out your survey and present the main findings in a report.
In your report you must include the following:
- What the survey is about and why you chose it as a topic to investigate.
- How you are going to sample the school population.
- How you trialled the questionnaire and what changes (if any) you made as a result and why.
- The main findings of your survey.
- Any weaknesses in your survey and how it could have been **improved**.
- Any recommendations for a future survey on the same issue.

You need to use:
- **characterising** skills to identify and describe properties of the data that you collect.
- **classifying** skills to organise the data into groups.
- **critiquing** skills to identify any problems or strengths with your survey.
- **improving** skills to refine your investigation.

 Now you have completed Unit 3, you may like to try the Unit 3 online knowledge test if you are using the Boost eBook.

Parallelograms, trapezia and circles

Trapezia is the plural of trapezium.

I.e. One trapezium and two trapezia.

- Use knowledge of rectangles, squares and triangles to derive the formulae for the area of parallelograms and trapezia.
- Use the formulae to calculate the area of parallelograms and trapezia.
- Understand π as the ratio between a circumference and a diameter.
- Know and use the formula for the circumference of a circle.

Parallelograms and trapezia

A parallelogram is a quadrilateral with two pairs of parallel sides. Here are two examples.

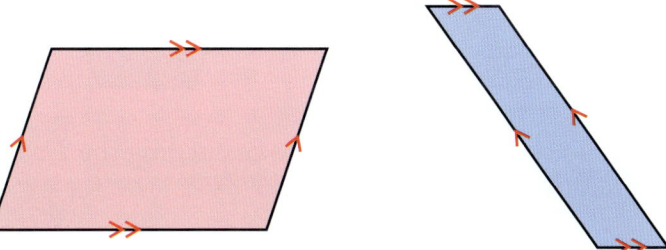

We use arrows to show that sides are parallel to one another.

To work out a rule for the area of a parallelogram, look at this parallelogram.

We can move the triangle on the right to the left-hand side of the parallelogram to make a rectangle.

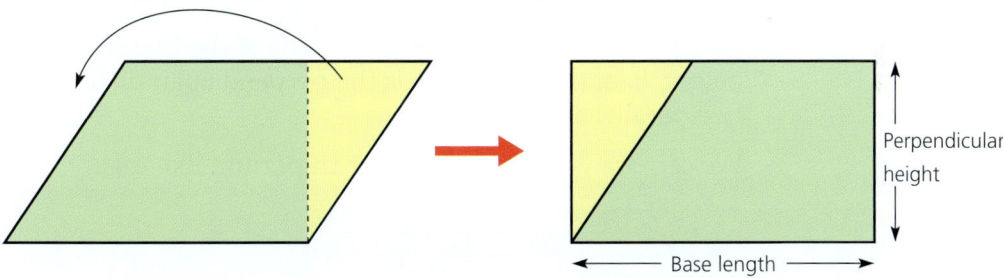

From the second diagram we can see that the base length and height of the parallelogram are exactly the same as those of the rectangle. Therefore, they must have the same area.

Area of a parallelogram = base length × perpendicular height

As with a triangle, the height must always be measured at right angles to the base.

A trapezium is also a quadrilateral. It has only one pair of parallel sides. Here are some examples.

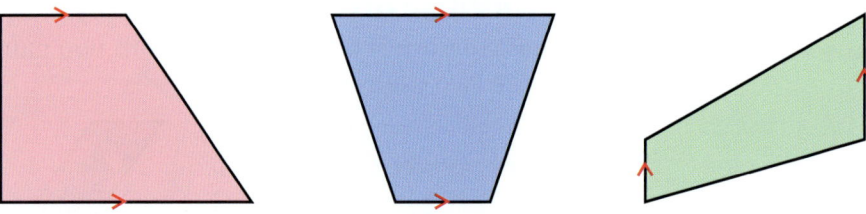

To work out a formula for calculating the area of a trapezium, look at this trapezium.

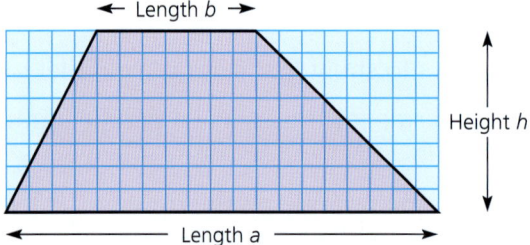

Length a and length b are the lengths of the two parallel sides. The height, h, of the trapezium is the perpendicular distance between the two parallel sides.

We can draw a rectangle on the trapezium like this.

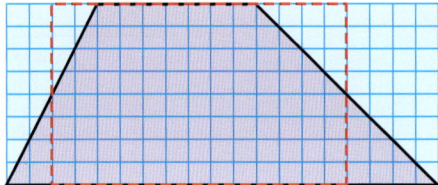

4 Parallelograms, trapezia and circles

We can show that the area of the trapezium is the same as the area of the rectangle.

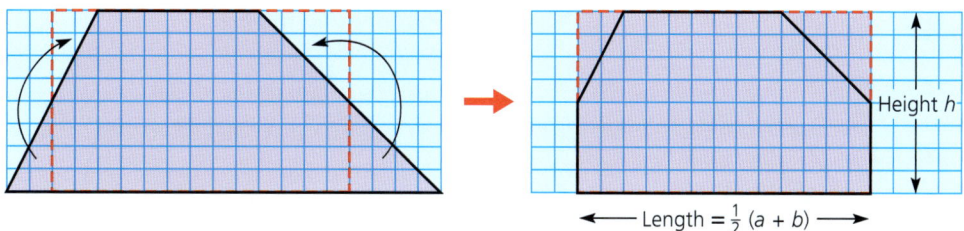

The height of the rectangle is the same as the height of the trapezium.

The length of the rectangle is the mean of the lengths of the two parallel sides.

The area of the trapezium is the same as the area of the rectangle.

Area of a trapezium $= \frac{1}{2}(a+b)h$

Worked example

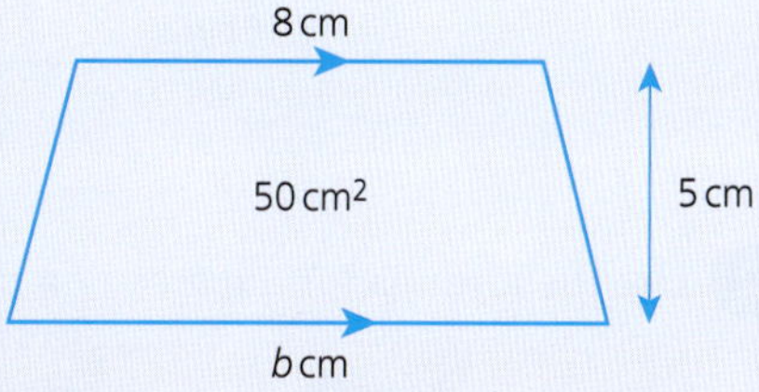

A trapezium is shown opposite.

Calculate the size of the side marked 'b'.

The formula for the area of a trapezium is $\frac{1}{2}(a+b)h$, where a and b are the lengths of the parallel sides and h the perpendicular distance between them. Substituting the values we know into the formula gives:

$50 = \frac{1}{2}(8+b) \times 5$

This can be rearranged, so we can deduce the value of b as follows:

$50 = 10 \times 5$

Therefore, $\frac{1}{2}(8+b) = 10$

Note that the writing in red must be worth the same.

As $\frac{1}{2}(8+b) = 10$, this can be written as $\frac{1}{2} \times 20 = 10$

Note that the writing in blue must be worth the same.

Therefore, $8 + b = 20$

So $b = 12$ cm

SECTION 1

Exercise 4.1

1 Calculate the area of each of these parallelograms and trapezia.

a

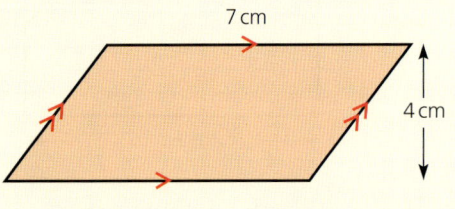

b

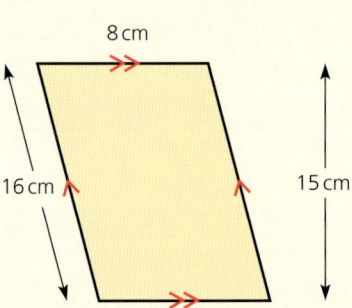

c

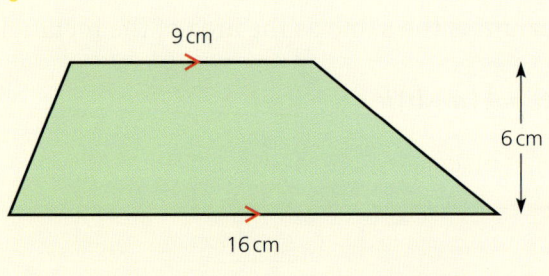

d

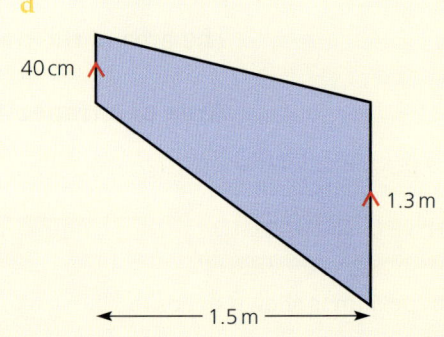

2 Use the formula for the area of a parallelogram to work out the missing values in this table.

	Length	Height	Area
a	4 cm	8 cm	
b	45 cm	0.2 m	
c	30 cm		150 cm²
d		0.45 m	45 cm²
e		250 cm	1 m²

4 Parallelograms, trapezia and circles

 3 Two different-shaped parallelograms, X and Y, are shown below. They share one side and fit between two parallel lines.

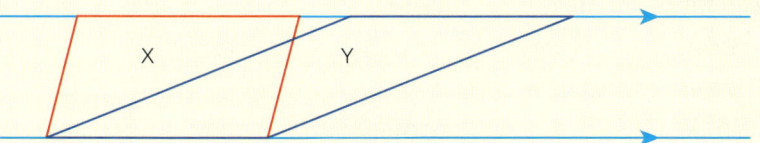

  a Which of the following statements is true?
  i) The area of X is greater than the area of Y.
  ii) The area of Y is greater than the area of X.
  iii) They both have the same area.
  iv) It is not possible to work out which one is bigger without knowing the lengths of the sides.
  b Give a **convincing** argument for your choice above.

 4 Look at the two congruent trapezia below.

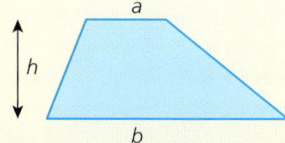 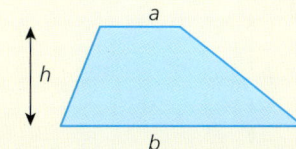

> Remember that congruent means exactly the same shape and size.

One of the trapezia is rotated 180° and joined to the other one as shown below

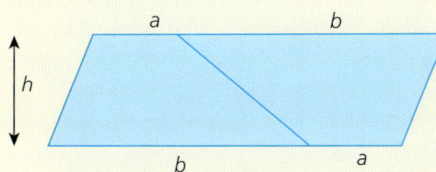

  a What shape has been made by joining the trapezia together?
  b Write an expression for the length of each of the longer sides.
  c Write an expression for the total area of the two trapezia.
  d Danya makes a **conjecture** about the formula for the area of a trapezium. His formula is $A = \frac{1}{2}(a+b)h$. By referring to your answer to part (c), explain why Danya's **conjecture** is correct.

SECTION 1

5 Use the formula for the area of a trapezium to work out the missing values in this table.

	Length a	Length b	Height	Area
a	3 cm	8 cm	4 cm	
b	20 cm	0.3 m	0.25 m	
c	10 cm	20 cm		500 cm²
d	0.25 m		40 cm	1100 cm²
e		45 cm	0.04 m	150 cm²

6 Calculate the area of each of these shapes.

a
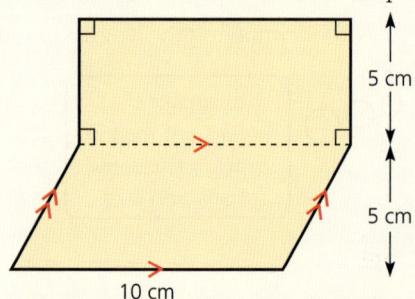

b
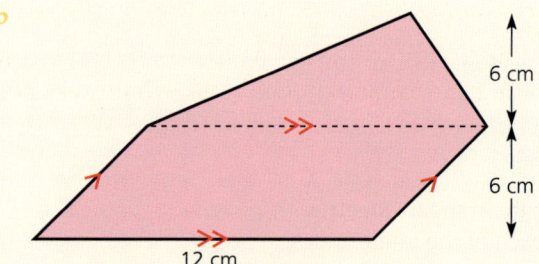

7 The area of the parallelogram in this diagram is 40 cm². The area of the triangle and the area of the parallelogram are equal. The area of the trapezium is twice the area of the triangle. Calculate the missing dimensions. Show your working clearly.

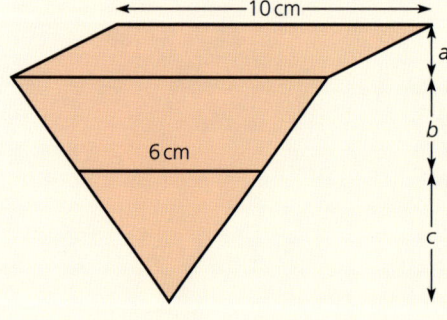

4 Parallelograms, trapezia and circles

8 This rectangular garden is split into four sections. Sections A and D are trapeziums; sections B and C are triangles. The area of section A is twice that of section B.

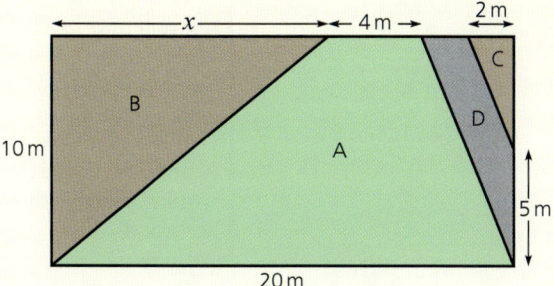

Showing your working clearly, calculate
a   the area of the whole garden
b   the area of section A
c   the length of the side marked x
d   the area of section D.

The circle

Circumference of a circle

You saw in Stage 7 that the perimeter of a circle is called its circumference; that is, the distance around the outside of the circle.

In order to calculate the circumference of a circle, you can use a formula. However, here is a simple practical task which will help you deduce an approximate formula for the circumference of a circle.

Practical activity

You will need
- a range of different-sized cylindrical objects
- a length of string
- a ruler
- paper and pencil (to record your results).

Method
- Use a ruler to measure the diameter of each cylinder.

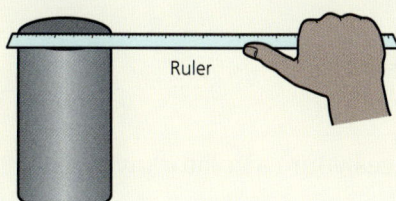

- Wrap string around each cylinder to measure its circumference.

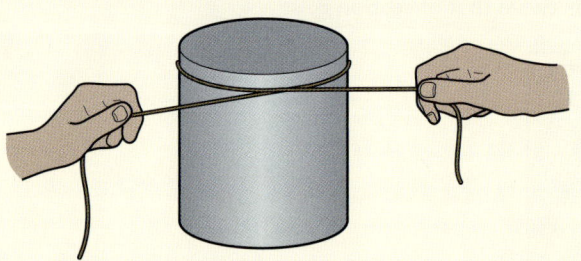

LET'S TALK

For thin cylinders, this method may produce large inaccuracies. Discuss with a partner why wrapping the string around the cylinder multiple times would **improve** the accuracy.

- Record your results for each cylinder in a table similar to this.

Cylindrical object	Circumference (cm)	Diameter (cm)

- Try to spot a pattern in your results, and write down a **conjecture** for a rule linking the length of the circumference to the length of the diameter.

Because of experimental errors, your rule will only be approximate.

You will have seen from your results in the activity that the length of the circumference of any circle is approximately three times the length of its diameter.

In fact, the circumference is always the diameter multiplied by a constant value known as pi (π).

Pi is not an exact number; it has an infinite number of decimal places.

To two decimal places, $\pi = 3.14$

To 14 decimal places, $\pi = 3.141\ 592\ 653\ 589$

Therefore, the circumference of any circle is given by the formula:

Circumference = $\pi \times$ diameter or $C = \pi D$

As the diameter is twice the radius, the circumference of a circle can also be given as:

Circumference = $\pi \times 2 \times$ radius or $C = 2\pi r$

Scientific calculators have a π key.

Check to see how many decimal places your calculator gives pi to.

Exercise 4.2

1 Calculate the circumference of each of these circles. The diameter of each circle has been given. Give your answers correct to two decimal places.

a) 6 cm

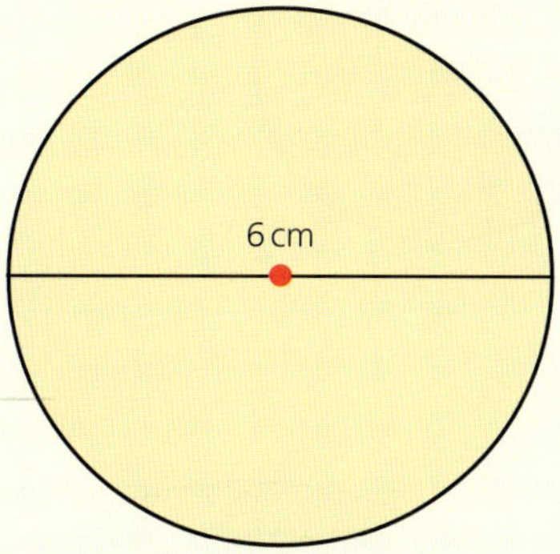

b) 25 cm

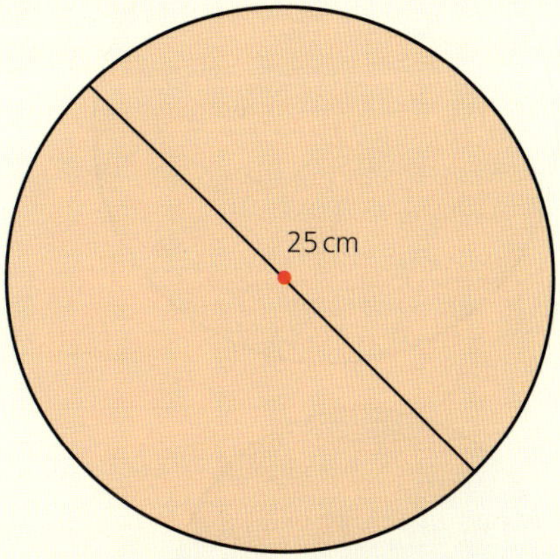

c) 40 mm

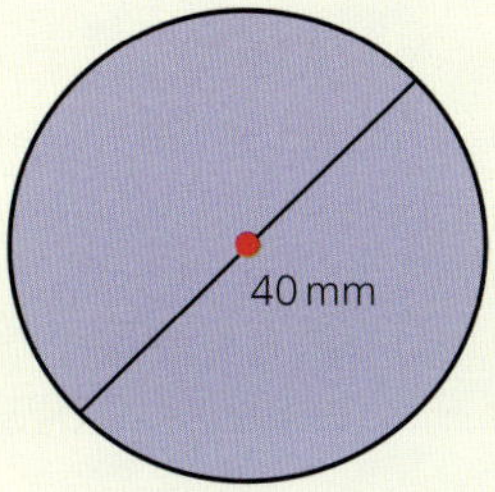

d) 1 m

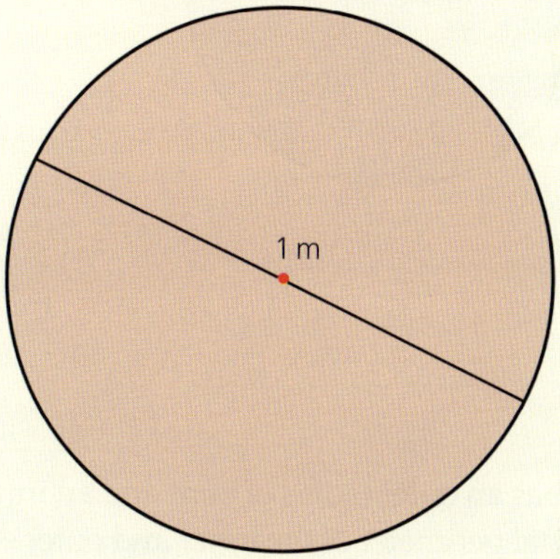

SECTION 1

2 Calculate the circumference of each of these circles. The radius of each circle has been given. Give your answers correct to two decimal places.

a 4 cm

b 3.5 cm

c 12 mm

d 6.3 m

3 A ring has an outer radius of 8 cm and an inner radius of r cm. If the total perimeter of the inner and outer edge of the ring is 62.8 cm (1 d.p.), calculate the length r, giving your answer as an integer.
Show all your working clearly.

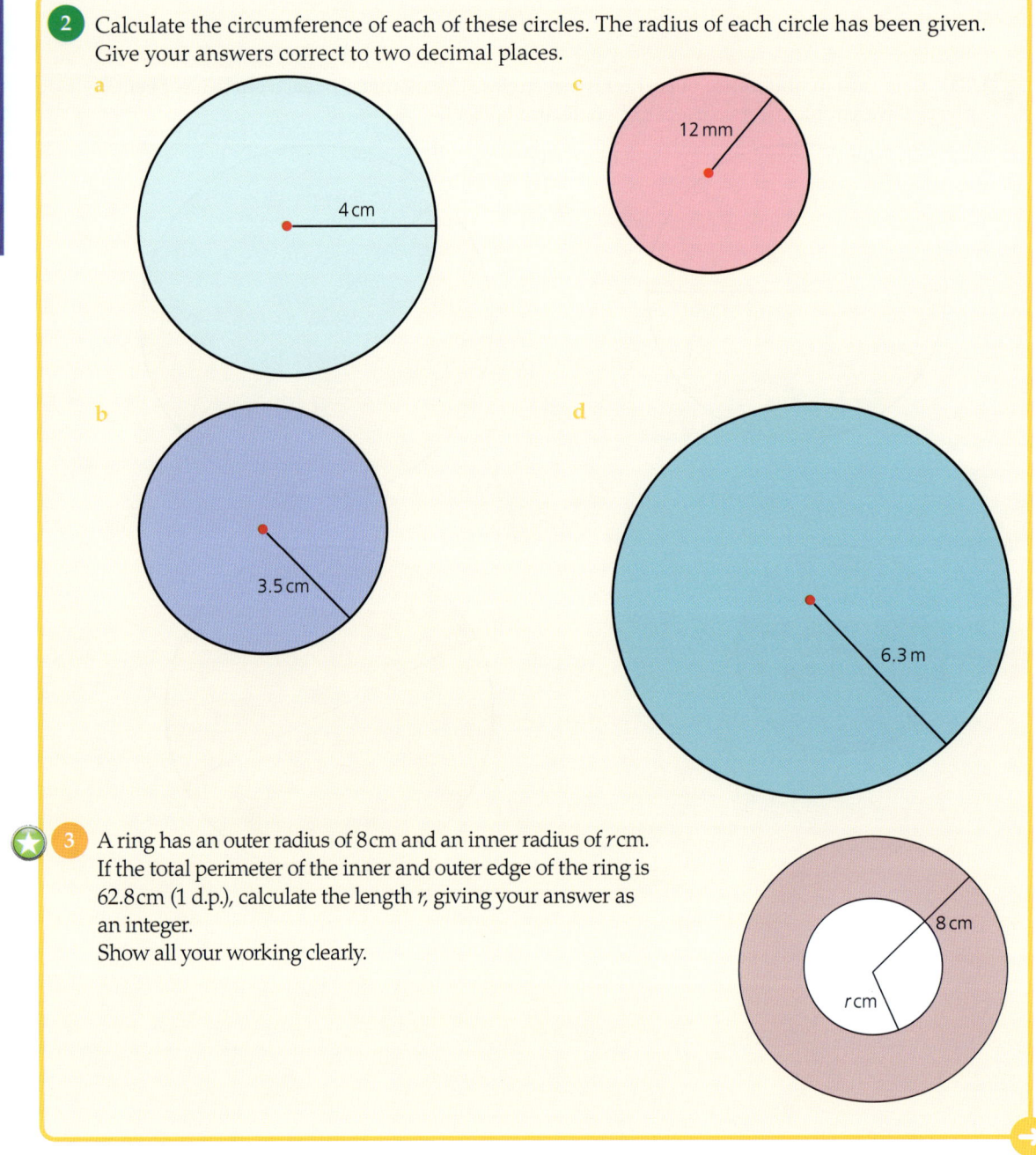

4 Parallelograms, trapezia and circles

⭐ **4** A circle when divided into four congruent parts forms four quadrants as shown.
Andrei states that the perimeter of a quadrant is simply a quarter of the circumference of a circle.
Show how you would **convince** Andrei that he is wrong.

A **quadrant** is a quarter of a circle.

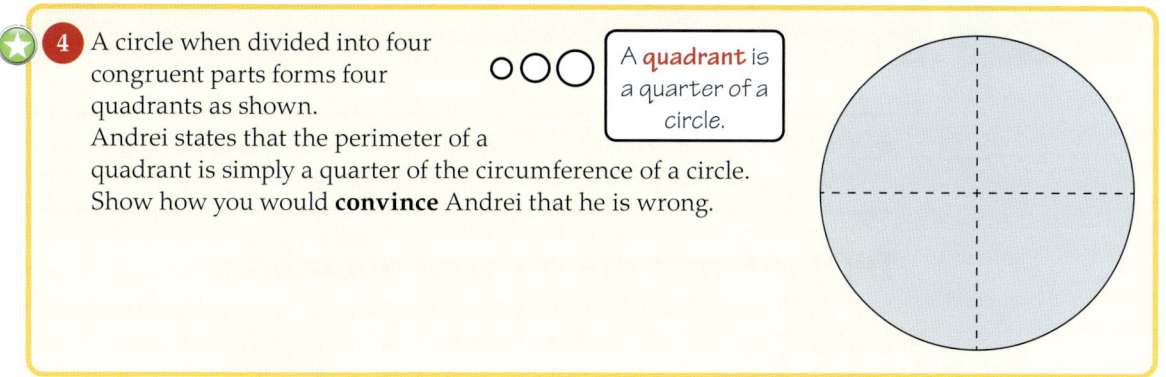

▶ Now you have completed Unit 4, you may like to try the Unit 4 online knowledge test if you are using the Boost eBook.

35

5 Order of operations

- Understand that brackets, indices (square and cube roots) and operations follow a particular order.

BIDMAS and the order of operations

You will remember from Stage 7 that although calculations that involve several operations are written on the page from left to right, they are not always done in that order. This is because in mathematics there is a hierarchy of operations. In other words, some operations are done before others.

Worked example

a Use a scientific calculator to work out the answer to this calculation.

$2 \times 4 + 5 =$

b Use a scientific calculator to work out the answer to this calculation.

$2 + 4 \times 5 =$

In the first example, doing the calculation from left to right gives the same answer as using a scientific calculator. In the second example, however, working from left to right would give an answer of 30, while a scientific calculator gives an answer of 22.

LET'S TALK
Why is it important to have an agreed order of operations?

The reason a scientific calculator does not work the calculation out from left to right is that, in mathematics, different operations are given different priorities; by convention, some types of operations are done before others. We can see from the two examples that multiplication is done before addition. Similarly, as subtraction is the opposite of addition, and division is the opposite of multiplication, division is done before subtraction. Therefore, multiplications and divisions in a calculation take priority over additions and subtractions.

5 Order of operations

Worked example

a Use a scientific calculator to work out the answer to this calculation.

$8 - 4 \div 4 =$

The answer is 7, and not 1, because the calculator works out $4 \div 4$ first and then subtracts the answer from 8.

b Use a scientific calculator to work out the answer to this calculation.

$9 + 7 \times 4 \div 2 =$

Here the multiplication and division are carried out first and then the answer is added to 9. **Note:** It does not matter which of multiplication or division is done first.

There will, however, be times when we will want an addition or subtraction to be done before a multiplication or division.

Consider the following calculation.

$6 + 8 \div 2 =$

A scientific calculator would work out the division first, giving a final answer of 10. However, if we wanted to work out $6 + 8$ first we would need to put this sum inside brackets. This is because brackets have the highest order of priority in any calculation, i.e. what is inside brackets is always done before any other operation.

Therefore:

while

Exercise 5.1

For questions 1–3, work out the answers
  i) mentally
  ii) using a scientific calculator.

1
a $12 \times 2 - 4$
b $8 \times 3 + 6$
c $4 + 10 \times 2$
d $16 - 8 \div 4$
e $44 - 4 \div 4$
f $9 \times 6 \div 3$

2
a $3 \times 2 + 4 \times 3$
b $8 + 3 \times 2 - 5$
c $9 - 6 \div 3 + 4$
d $20 \div 5 \times 2 - 8$
e $16 - 8 \div 8 + 3 \times 4$
f $5 + 3 \times 2 \div 6 - 5$

SECTION 1

3
a $(4+1) \times 3$
b $6 \times (8-5)$
c $(24-6) \div 9$
d $(3+2) \times (12-5)$
e $(4-1)+3 \times 4$
f $6 \times (3+5) \div 8$

For questions 4–6:
i) copy the calculation and put in any brackets that are needed to make it correct
ii) check your answers using a scientific calculator.

4
a $12-8 \times 2 = 8$
b $5 \times 2 + 4 = 30$
c $2 \times 3 + 4 - 5 = 4$
d $10 - 4 \times 3 + 3 = 36$
e $9 + 6 - 3 \div 2 + 4 = 10$
f $9 + 6 - 3 \div 2 + 4 = 2$

5
a $20 - 8 \div 2 + 6 = 22$
b $20 - 8 \div 2 + 6 = 12$
c $20 - 8 \div 2 + 6 = 1.5$
d $20 - 8 \div 2 + 6 = 10$
e $20 - 8 \div 2 + 6 = 19$

6
a $8 + 3 \times 4 - 6 = 14$
b $8 + 3 \times 4 - 6 = 38$
c $8 + 3 \times 4 - 6 = -22$
d $8 + 3 \times 4 - 6 = 2$

> The 'to the power of' buttons are all examples of indices.

Indices and roots

Calculators have many different keys, some of which are simply more efficient ways of carrying out ordinary calculations.

Examples of this are the 'to the power of' buttons.

You will already be familiar with some of them.

LET'S TALK

'Squaring' is the same as raising a number to the power of 2.

What is the term for raising a number to the power of 3?

Worked example

a Using a calculator work out 43×43.

43×43 can also be written as 43^2. This is forty-three squared or 43 to the power of 2.

The calculator provides an efficient way of squaring a number by using the $\boxed{x^2}$ key.

Therefore, 43^2 can be typed into the calculator as:

$\boxed{43}\,\boxed{x^2}\,\boxed{=}\,\boxed{1849}$

b Using a calculator, work out $6 \times 6 \times 6 \times 6$.

The calculation could be typed into the calculator as $6 \times 6 \times 6 \times 6$, but this is not efficient. $6 \times 6 \times 6 \times 6$ can also be written as 6^4, i.e. 6 to the power of 4.

On most calculators the $\boxed{y^x}$ key or the $\boxed{\wedge}$ key is the 'to the power of' or index key.

So 6^4 can be entered into the calculator as:

KEY INFORMATION

The word 'index' refers to the power of a number. The plural of index is indices.

5 Order of operations

Exercise 5.2

1. Explain what the √ button does on your calculator.

2. Using a calculator work out the following in parts (a)–(c)
   a $\sqrt{16}$
   b $\sqrt{36}$
   c $\sqrt{100}$

3. Using a calculator work out the following in parts (a)–(c)
   a $\sqrt[3]{27}$
   b $\sqrt[3]{125}$
   c $\sqrt[3]{1728}$

4. Kwasi and Sayo are discussing the results of squaring and cubing numbers. They conclude that cubing a number will always give a bigger answer than squaring that number.
   a Decide whether this is
      • always true
      • sometimes true
      • never true.
   b Explain your choice of answer above. Give examples to support your answer.

   > You need to use your **specialising** skills and check different types of numbers.

Worked example

a i) Using a calculator work out 2×4^3.

ii) Decide on the order in which the calculation has been done.

The 4^3 has been done first to give 64 and then that has been multiplied by 2 to give an answer of 128.

In this calculation it is important to realise another priority in calculations.

Indices are second in order of priority, after brackets.

Therefore, in the calculation above, 4^3 is worked out before the multiplication by 2.

Note: If we wanted to multiply the 4 by the 2 first, before raising it to the power of 3, we would need to put the 2×4 in brackets, i.e. $(2 \times 4)^3$.

SECTION 1

b Zac works out the following calculation on paper as follows:

$$35 - 20 \times \sqrt[3]{64} = 35 - 20 \times 4$$
$$= 15 \times 4$$
$$= 60$$

He checks the answer using a calculator but gets a different answer.

i) Where has he gone wrong?

He correctly worked out $\sqrt[3]{64}$ first to give a value of 4.

However, in the next line of calculation he should have done the 20×4, but instead he has done the $35 - 20$ first. Multiplications should be done before subtractions.

ii) What should the correct answer be?

$$35 - 20 \times \sqrt[3]{64} = 35 - 20 \times 4$$
$$= 35 - 80$$
$$= -45$$

> You can use the shorthand 'BIDMAS' to help remember this.

Therefore, the order of priority in calculations is:
1. **B**rackets
2. **I**ndices (including square and cube roots)
3. **D**ivision and/or **M**ultiplication
4. **A**ddition and/or **S**ubtraction.

Exercise 5.3

1 Using a scientific calculator, work out the answers to the following calculations.

a $3^2 + 4$
b $8^2 - 3^3$
c $\frac{6^3}{16+2}$
d $\frac{(5+3)^2}{6-1}$
e $\frac{4^3 + 6^2}{2^5} - 3$
f $\frac{(4+3)^3}{14} - 24$
g $\sqrt{26 + 10} + 3^2 \times 4$
h $\frac{8 \times \sqrt[3]{150 - 25}}{4^2}$
i $4 + \sqrt{49} \times \sqrt[3]{27}$

2 i) Ade says that each of the following pairs of calculations have the same answer. Is Ade correct? Explain your answers fully.

ii) Calculate the answer to each calculation.

a $\sqrt{7 + 3^2}$ and $\sqrt{(7+3)^2}$
b $3^2 + \frac{15}{3}$ and $\frac{3^2 + 15}{3}$
c $\sqrt{\frac{81}{9}} + 4^2$ and $\frac{\sqrt{81}}{9} + 4^2$

> Note that $\frac{3^2 + 15}{3}$ can also be written as $\frac{(3^2 + 15)}{3}$. But it is more common to write it without the brackets.

Now you have completed Unit 5, you may like to try the Unit 5 online knowledge test if you are using the Boost eBook.

Expressions, formulae and equations

- Understand that letters have different meanings in expressions, formulae and equations.
- Understand that the laws of arithmetic and order of operations apply to algebraic terms and expressions.

You will already be familiar with the fact that in algebra you are dealing with letters as well as numbers.

These letters, because they can take on different values are known as variables. The numbers, because their value is fixed, are known as constants.

Work involving algebra can take many forms. We will be covering the three most common types:

	Expression	Formula	Equation
e.g.	$2(a+b)$	$v = u + at$	$3x + 2 = 11$

The meaning of the letters depends on the different form the algebra takes.

This unit will look at these different meanings.

LET'S TALK
From the examples given here, what do you think are the differences between an expression, a formula and an equation?

Expressions

An algebraic **expression** is simply a statement written using algebra. It will contain at least one **variable** and may contain **constants** and **mathematical operations**.

Mathematical operations are things like +, −, × and ÷ but can also include others such as $\sqrt{}$ and $\sqrt[3]{}$

e.g. $3x + y$ $\quad \frac{x^2 + 5}{3} \quad$ $\sqrt[3]{(3p + 2q)}$

In the expression $3x + y$ above there are two **terms**: '$3x$' and 'y'.

The term 'y' still has a coefficient. It has a coefficient of 1, but it is not usually included.

The number multiplying a variable is known as the **coefficient** of that variable.

In the example above in the term $3x$ the '3' is the coefficient of x.

Expressions often will represent a mathematical situation.

41

SECTION 1

Worked example

A cuboid has dimensions as shown.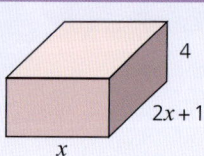

a Write an expression for the volume of the cuboid.
   $4x(2x+1)$

b Which part of the expression is a variable?
   x is the variable as it can take different values.

c Work out the value of the expression $4x(2x+1)$ when $x = 6$.
   When calculating using algebra, the order of operations still follows the rules of BIDMAS.
   Therefore when $x = 6$
   $$4x(2x+1) = 4 \times 6 \times 13$$
   $$= 312$$

> Note that $4x(2x+1)$ means 4 times x times $(2x+1)$.
>
> As it does not have an '=' sign it is an expression. If it had been written as $V = 4x(2x+1)$ then it would be considered a formula.

Exercise 6.1

1. Identify which are expressions in the following:
   a $A = \frac{1}{2}bh$
   b $2(w + l)$
   c $2x + 6 = 9$
   d $10p + 15$

2. Substitute the following values into the expressions below to work out their value.
   $a = 2, b = 3, c = 4, d = 5$
   a $3a + b$
   b $ab - cd$
   c $a(b+d)^2$
   d $\dfrac{b^3 - 2c - d}{a}$
   e $d - \dfrac{a^2}{c}$

3. Write each of the following as an expression:
   a 3 is added to x and the result squared
   b the product of x and y is cubed and the result divided by 4
   c p is squared and subtracted from q squared. The result is divided by 2.

> The product of two values means to multiply them together.

6 Expressions, formulae and equations

4. A stick of length $2x$ cm has a length of 5 cm cut from it.
   a. Write an expression for the new length of the stick.
   b. How many of the new sticks would need to be joined together to form a length of $(8x - 20)$ cm? Justify your answer.

5. A cuboid has dimensions as shown.
   a. Explain why the expression for the total surface area is given by $2m^2 + 4mn$.
   b. Calculate the value of the total surface area when $m = 5$ cm and $n = 8$ cm.

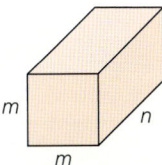

6. A rectangle of length a cm and width b cm has a square of side length c cm cut from it.
   a. Write an expression for the area of the shaded shape that is left.
   b. If the shaded area has a value of 28 cm², give a possible value for each of the lengths a, b and c.

7. A cube has a volume of $27t^3$.
   a. Write an expression for the length of each side of the cube.
   b. Calculate the volume of the cube when $t = 2$ cm.
   c. What does the expression $9t^2$ represent for this cube? Justify your answer.

LET'S TALK

Is there more than one combination for the values of a, b and c? Justify your answer.

Can they have decimal values? What strategies can you use to work out possible values?

Equations

An **equation** can also be written using algebra and numbers. It is different from an expression as it will always have an equals sign, whereas an expression will not.

The equals sign implies that what is on one side of the equation will be worth the same as what is written on the other side.

Examples include.

$$2x - 4 = 8 \qquad 6a - 4 = 5 - 3a \qquad y^2 - 4 = 12$$

In each of the equations above you can see that there is one variable. As they are equations, what is on the left side is equal to what is on the right side, but there is an unknown quantity (i.e. x, a or y). To find out the value of that unknown quantity the equation needs to be **solved**. This is done by **rearranging** it.

Note: Although the unknown in an equation is an algebraic term, it is different from a variable in an expression. In an expression the variable can take any value, and this will affect the size of the expression. In an

LET'S TALK

Can you find the value of the unknown in each case that makes the left-hand side equal to the right-hand side?

SECTION 1

equation, the unknown can only have specific value(s). These value(s) have to make the left-hand side of the equation equal to the right-hand side.

> Just saying 'Solve the equation' implies having to find the value of the unknown quantity.

Worked example

a Solve the equation $2x + 4 = 12$ to find the value of the unknown.

Remember, to keep the equation balanced what we decide to do to one side of the equation, we must also do to the other.

$2x + 4 = 12$

$\quad 2x = 8$ (4 subtracted from both sides)

$\quad\ x = 4$ (both sides divided by 2)

The answer can be checked by substituting it back into the equation

$2 \times 4 + 4 = 12$

b Solve the equation $12 - 5x = 2$.

It is often easier to work with a positive unknown quantity rather than with a negative one.

$12 - 5x = 2$

$12 = 2 + 5x$ (5x added to both sides)

$10 = 5x$ (2 subtracted from both sides)

$\ 2 = x$ (Both sides divided by 5)

$\ x = 2$

c A line's length is given by the expression $4x - 6$.

If the line is 20 cm long, form an equation and solve it.

If the line is both $4x - 6$ and 20 then they must be equal to each other.

$4x - 6 = 20$

$\quad 4x = 26$ (add 6 to both sides)

$\quad\ x = 6.5$ (divide both sides by 4)

Exercise 6.2

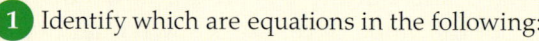

1 Identify which are equations in the following:
   a $A = \frac{1}{2}bh$
   b $2(w + l)$
   c $2x + 6 = 9$
   d $10p + 15 = 30$

6 Expressions, formulae and equations

2 i) Solve the following equations to calculate the unknown quantity.
   ii) Check your answers each time by substituting them back into the equations.
   a $3x - 9 = 12$
   b $7 + 4x = 11$
   c $8 - 6x = 0$
   d $15 = -x + 20$
   e $2 = \frac{1}{2}x - 1$

3 If $x = 5$, which of the following equations are correct? Show your reasoning clearly.
   $5 + x = 10$ $3 + x^2 = 28$ $4x - 6 = 12$
   $\sqrt{3x + 10} = x$ $x^3 + x^2 - 150 = 0$ $2(4x)^2 = 200$

4 Sara is solving an equation. She has made a mistake.
   $6 + 3x = 24$
   $6 + x = 8$ (divide both sides by 3)
   $x = 2$ (subtract 6 from both sides)
   a Explain what mistake she has made.
   b Solve the equation correctly.

5 A rectangle has dimensions as shown. The perimeter of the rectangle is 74 cm.
   a Write an equation from the information given.
   b Solve the equation.
   c Calculate the area of the rectangle.

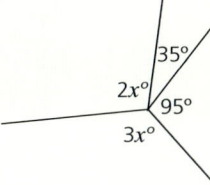
$x + 2$ cm
$3x - 1$ cm

LET'S TALK
How could this problem be solved using ratios?

6 Four angles are arranged around a point as shown.
   a By forming an equation, calculate the value of x.
   b Calculate the size of each of the two unknown angles.

$35°$
$2x°$
$95°$
$3x°$

7 The cuboid has dimensions as shown. The cuboid has a volume of 24 cm³.
   a Find the value of x.
   b Calculate the total surface area of the cuboid.

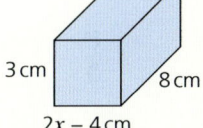
3 cm
8 cm
$2x - 4$ cm

Formulae

Formulae is the plural of formula. I.e. 1 formula and 2 formulae

A formula looks similar to an equation in that it has an equals sign. However, it also has differences. An equation is used to find a value of an unknown variable. Only certain values will make the equation work.

For example, the equation $x + 6 = 10$ is only correct if $x = 4$.

A formula, however, shows the relationship between different variables. You will already be familiar with some formulae.

SECTION 1

For example, the formula for the area of a parallelogram is $A = lh$,

where A represents the area, l, its length and h its height.

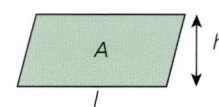

Here there are three variables, A, l and h.

As A is on its own on one side it is called the subject of the formula. To work out A the values of the other two variables need to be given and these can take any value.

If the area and height are known and the length needed, then the formula needs to be rearranged to make l the subject. This is done in the same way as when working with an equation, i.e. what is done to one side of the formula must also be done to the other side.

> **Worked example**
>
> a The formula for finding the area of a triangle is $A = \frac{1}{2}bh$ where A is the area, b is the length of the base and h, its perpendicular height.
>
>   i) Calculate the area of a triangle when $b = 20$ cm and $h = 35$ cm.
>
> $A = \frac{1}{2}bh$
>
> $A = \frac{1}{2} \times 20 \times 35$
>
> $A = 10 \times 35$
>
> $A = 350$ cm²
>
>   ii) Rearrange the formula to make h the subject and calculate the height if $A = 100$ cm² and $b = 25$ cm.
>
> $A = \frac{1}{2}bh$
>
> $2A = bh$ (multiply both sides by 2)
>
> $\frac{2A}{b} = h$ (divide both sides by b)
>
> $h = 2A/b$
>
>   Substituting the values of A and b gives:
>
> $h = \frac{2 \times 100}{25} = 8$
>
>   Therefore $h = 8$ cm

6 Expressions, formulae and equations

Exercise 6.3

1. **Classify** the following into expressions, equations and formulae.
   a $\sqrt{a^2 + b^2}$
   b $y = 3x - 4$
   c $v = u + at$
   d $9x - 2 = 16$
   e $x^2 = 36$
   f $C = 2\pi r$
   g $p^2 + p - 3$

2. a Write down three different formulae that you can remember from your earlier work in mathematics.
   b Explain what each of the formulae are used for and what each of the different variables represent.

3. The formula for the circumference (C) of a circle is given by $C = \pi D$, where D represents the diameter of the circle.
   a What are the variables in this formula?
   b Are there any constants? Justify your answer.
   c Calculate the circumference of a circle with a diameter of 12 cm.
   d i) Rearrange the formula to make D the subject.
      ii) Use your formula to calculate the diameter of a circle with a circumference of 45 cm. Give your answer to 1 decimal place.

4. An approximate formula used for converting a temperature in degrees Fahrenheit (F) to degrees Centigrade (C) is given as, $C = \frac{1}{2}(F - 32)$.
   a Use the formula to convert
      i) 32 °F to Centigrade
      ii) 100 °F to Centigrade.
   b Rearrange the formula to make F the subject by following these steps:
      • Double both sides of the formula.
      • Add 32 to both sides.
   c The hottest temperature recorded on Earth was 56.7 °C in Death Valley, California in 1913. Use your formula in part (b) to convert the temperature to degrees Fahrenheit.

5. In a branch of mathematics called kinematics, there is a formula to calculate the final velocity of an object. It is $v = u + at$, where v is the final velocity, u, its starting velocity, a, its acceleration and t, the time.
   a Calculate the value of v when $u = 5$, $a = 4$ and $t = 12$.
   b i) Rearrange the formula to make t the subject.
      ii) Use your formula to calculate t when $v = 20$, $u = 6$ and $a = 0.5$.

LET'S TALK

What are the hottest and coldest recorded temperatures in your country? Write down the figures in both °F and °C.

How accurate are these formulae when converting from one to the other?

Kinematics is a branch of mathematics that deals with the motion of objects.

Now you have completed Unit 6, you may like to try the Unit 6 online knowledge test if you are using the Boost eBook.

7 Recording, organising and representing data

- Record, organise and represent categorical, discrete and continuous data.
- Choose and explain which representation to use in a given situation.

You will already be familiar with many of the types of graph presented here. This unit will cover the different types in more depth but also focuses on what type of graph to choose for the type of data collected.

Once the data have been collected, they can be displayed in many ways. The methods available will depend on the type of data collected.

Two-way tables

Two-way tables can be used when data are split into two main categories. Here the data regarding the year are **discrete** and the data about gender are **categorical**.

		\multicolumn{5}{c}{Year}
		7	8	9	10	11
Gender	Girls	25	31	19	34	37
	Boys	20	32	23	30	40

For example, this two-way table shows the numbers of girls and boys in each of Years 7–11 in a school.

The two main categories here are year group and gender, but each of these categories has got sub-divisions. The information can be read off fairly easily. From the table we can see that there are 32 boys in Year 8. The total number of students in Year 11 can be calculated as 77.

Venn diagrams

Venn diagrams are used to show the frequency of data in two or three categories.

The Venn diagram on the next page shows the number of students in a class using social media applications, *F* and *I*.

> **KEY INFORMATION**
> Carroll diagrams are a special type of two-way table.

> **LET'S TALK**
> Can the categories also be for continuous data?

> **LET'S TALK**
> What other types of graph could show the data clearly?

7 Recording, organising and representing data

LET'S TALK

Could there be a Venn diagram with three rings? Give an example if you think it's possible.

If the rings did not intersect what would that mean?

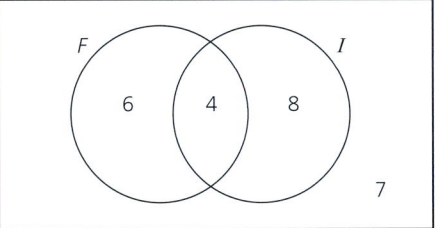

The total of all the numbers gives the number of students in the survey. Six students used only application F, eight only used application I, four used both applications, while seven did not use either.

LET'S TALK

How can you show this information on a Carroll diagram?

Frequency tables and diagrams

Frequency tables are a popular way of displaying data. If the data are discrete, then the information in the table can be displayed using a frequency diagram. For example, the frequency table and bar chart below show the number of pets in each house in a particular street.

Number of pets	Frequency
0	12
1	15
2	10
3	4
4	1
more than 4	2

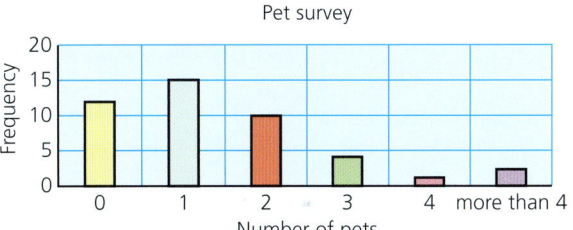

Bar charts

In Stage 7 you encountered both dual frequency diagrams and bar graphs. These can be combined in a compound bar chart to compare two sets of data on the same frequency diagram.

For example, 30 students in a class were asked how they travel to school. The results are shown in the table below according to gender.

	Car	Bicycle	Walk	Bus	Other
Girls	5	2	4	3	2
Boys	6	4	1	3	0

49

SECTION 1

As a dual bar chart, the information can be displayed as shown in the figure below:

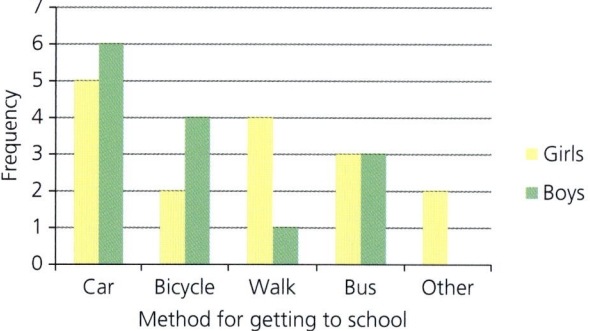

Alternatively the same data could be displayed using a compound bar chart as shown in the figure below.

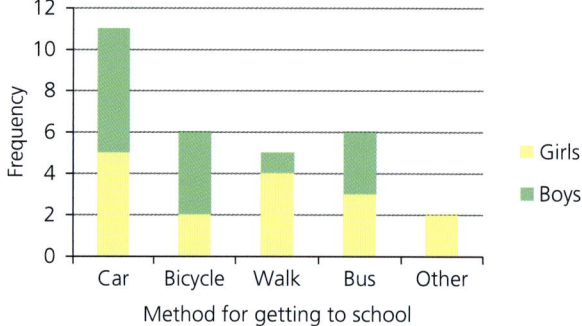

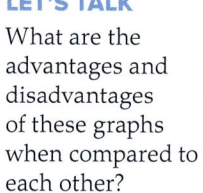

LET'S TALK
What are the advantages and disadvantages of these graphs when compared to each other?

Grouped frequency diagrams

A grouped frequency diagram displays the frequency of either continuous or grouped discrete data in the form of bars. There are several important features of a grouped frequency diagram:
- The bars are only joined together for continuous data.
- The bars all have the same width.
- The frequency of the data is represented by the height of each bar.

7 Recording, organising and representing data

Worked example

This table shows the marks out of 100 in a mathematics test for a class of 32 students. Draw a grouped frequency diagram to represent these data.

Score	Frequency
1–10	0
11–20	0
21–30	1
31–40	2
41–50	5
51–60	8
61–70	7
71–80	6
81–90	2
91–100	1

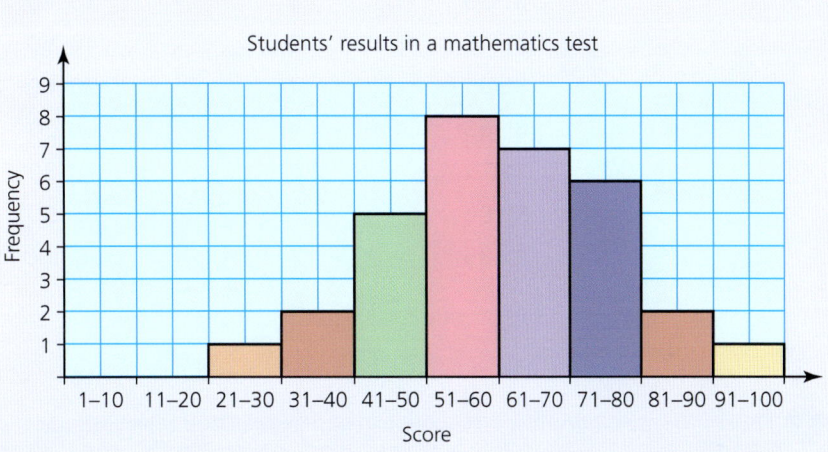

All the class intervals are the same. As a result, the bars of the grouped frequency diagram will be of equal width, and frequency can be plotted on the vertical axis.

Notice how the groups do not overlap. They represent grouped discrete data.

In the worked example above, groups of 10 marks were used. The group size that is chosen will affect the shape of the graph. If larger groups are used, accuracy is lost as each group will contain a lot of data. For example

Score	Frequency
1–50	8
51–100	24

However, groups that are too small, although more accurate, may make the graph difficult to analyse as most groups will contain very little data. Look at the following table, for example.

SECTION 1

Score	Frequency	Score	Frequency
1–5	0	6–10	0
11–15	0	16–20	0
21–25	0	26–30	1
31–35	1	36–40	1
41–45	2	46–50	3
51–55	4	56–60	4
61–65	4	66–70	3
71–75	3	76–80	3
81–85	1	86–90	1
91–95	1	96–100	0

Exercise 7.1

 1 This two-way table shows the favourite subjects of 100 boys and girls.

		Subject				
		Mathematics	Science	Sport	Languages	Other
Gender	Boys	12	6	17	7	2
	Girls	21	12	6	14	3

a Which is the most popular subject among the girls?
b Which is the most popular subject among the boys?
c How many girls chose science as their favourite subject?
d How many students chose languages as their favourite subject?
e How many girls were included in the survey?
f   i) Present these data using a graph of your choice.
    ii) Give a **convincing** reason for your choice of graph.

LET'S TALK
Compare your infographic to those of other students in your class. Discuss what makes one infographic more effective than another.

2 The numbers of litres of milk consumed in a group of 150 houses are shown in the table.

Number of litres	1	2	3	4	5	6
Frequency	27	54	34	16	15	4

a Show this information on a frequency diagram.
b Design your own infographic to display these data.
c Comment on the effectiveness of your infographic. How could you **improve** it?

LET'S TALK
Decide which of these two diagrams is the most effective way to display the data. Give reasons for your answer.

7 Recording, organising and representing data

3 A large firm records the number of computer repairs carried out each day over a 31-day period. The results are shown below.

16	37	43	62	52	19	43	38	23	17	9
47	64	46	43	16	48	16	38	27	53	
44	34	25	52	39	39	18	15	12	8	

  a Make a grouped frequency table with class intervals 0–9, 10–19, 20–29 etc.
  b Illustrate the data on a grouped frequency diagram.

4 The masses of 50 rugby players attending a tournament are recorded in this grouped frequency table.

Mass (kg)	Frequency
$70 \leq M < 80$	3
$80 \leq M < 90$	7
$90 \leq M < 100$	10
$100 \leq M < 110$	20
$110 \leq M < 120$	7
$120 \leq M < 130$	3

> **LET'S TALK**
> What type of data is the mass? Why are the groups written in this way?

  a Illustrate this information on a grouped frequency diagram.
  b What other type of graph could be used to illustrate these data? Draw a sketch of what the graph would look like.

5 This table shows the distances travelled to school by a class of 32 students.

Distance (km)	Frequency
$0 \leq d < 1$	9
$1 \leq d < 2$	4
$2 \leq d < 3$	7
$3 \leq d < 4$	2
$4 \leq d < 5$	5
$5 \leq d < 6$	1
$6 \leq d < 7$	3
$7 \leq d < 8$	1

  a Illustrate this information on a grouped frequency diagram.
  b Carry out a similar survey on students in your class.
   i) Decide what group size to use and produce a grouped frequency table of your data.
   ii) Plot a grouped frequency diagram of your data.
   iii) Comment on your results.

SECTION 1

 6 These data record the percentage scores of students in a mathematics exam.

30	47	43	58	62	73	47	59	68	51	57
64	66	70	36	60	57	83	64	61	41	58
58	72	67	88	56	58	87	70	40	57	56
64	70	48	62	73	67	69	58	80	74	59

a i) Choose a suitable class interval and make a grouped frequency table of the data.
   ii) Choose a different class interval from that in part (i) and complete another grouped frequency table of the data.
b i) Justify which of the class intervals you looked at in part (a) above is most suitable for the data.
   ii) Plot a grouped frequency diagram of the data using the class interval you decided on.
   iii) By referring to your grouped frequency diagram, make one conclusion about the results.

7 a Collect some test data from your class, separating the data for boys and girls. Present the data in a grouped frequency table choosing suitable class intervals.
   b Display your data in:
     i) a dual frequency diagram
     ii) a compound frequency diagram.
   c Write one conclusion you can make from each diagram, which cannot be easily made from the other diagram.

Pie charts

Pie charts are another popular way of displaying data. With a pie chart, each **sector** (slice) represents a fraction of the total. Its size is proportional to the frequency of that category as a fraction of the total. They can be used for discrete, continuous or categorical data.

For example, this table shows the numbers of different flavours of ice creams sold, and their fractions of the total number sold.

Flavour	Frequency	Fraction of the total
Vanilla	25	$\frac{25}{100} = \frac{1}{4}$
Chocolate	50	$\frac{50}{100} = \frac{1}{2}$
Strawberry	25	$\frac{25}{100} = \frac{1}{4}$

7 Recording, organising and representing data

Converting these data to a pie chart is relatively straightforward, as the circle can be split into the fractions fairly easily.

Ice cream flavours

If the numbers are not so straightforward, a pie chart scale (marked in percentages) or an angle measurer or protractor (marked in degrees) must be used to construct the pie chart.

Worked example

The table below shows the numbers of brothers and sisters of 30 students in a class.

Number of brothers and sisters	0	1	2	3	4	5
Frequency	5	8	11	3	2	1

Show this information on a pie chart.

To express these figures as **percentages**, we need to work out what fraction each frequency is compared with the total and then multiply it by 100.

Number of brothers and sisters	Frequency	Fraction of the total	Percentage
0	5	$\frac{5}{30}$	$\frac{5}{30} \times 100 = 17\%$
1	8	$\frac{8}{30}$	$\frac{8}{30} \times 100 = 27\%$
2	11	$\frac{11}{30}$	$\frac{11}{30} \times 100 = 37\%$
3	3	$\frac{3}{30}$	$\frac{3}{30} \times 100 = 10\%$
4	2	$\frac{2}{30}$	$\frac{2}{30} \times 100 = 7\%$
5	1	$\frac{1}{30}$	$\frac{1}{30} \times 100 = 3\%$

The pie chart would then be drawn using a **pie chart scale**.

To express these figures as **angles**, we need to work out what fraction each frequency is compared with the total and then multiply it by 360 (as there are 360° in a full circle).

SECTION 1

Number of brothers and sisters	Frequency	Fraction of the total	Angle
0	5	$\frac{5}{30}$	$\frac{5}{30} \times 360 = 60°$
1	8	$\frac{8}{30}$	$\frac{8}{30} \times 360 = 96°$
2	11	$\frac{11}{30}$	$\frac{11}{30} \times 360 = 132°$
3	3	$\frac{3}{30}$	$\frac{3}{30} \times 360 = 36°$
4	2	$\frac{2}{30}$	$\frac{2}{30} \times 360 = 24°$
5	1	$\frac{1}{30}$	$\frac{1}{30} \times 360 = 12°$

The pie chart would then be drawn using an angle measurer or **protractor**.

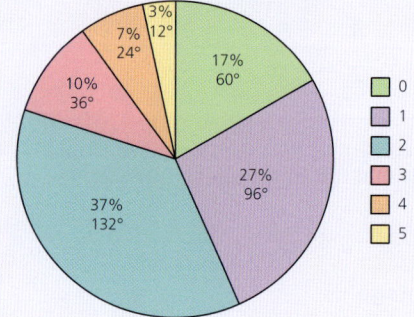

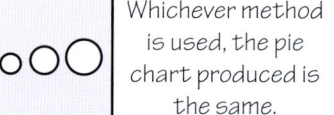

Whichever method is used, the pie chart produced is the same.

LET'S TALK
In this example why do the percentages in the pie chart total 101% rather than 100%?

Exercise 7.2

1. 90 students sat a mathematics exam. On the way out of the hall, they were asked to say whether they had found it hard, OK or easy. Here are the results. Show the results on a pie chart.

	Response			
	Easy	OK	Hard	No reply
Frequency	22	37	19	12

2. Two football teams have the results shown in the table. Illustrate these on two pie charts.

	Total	Win	Draw	Lose
Spain	36	22	8	6
Turkey	36	21	13	2

7 Recording, organising and representing data

3 a Write a questionnaire to find the favourite subject of a group of students in your class and display the results on a pie chart.
b Write a brief conclusion about the results of your questionnaire. How could you **improve** your questionnaire?

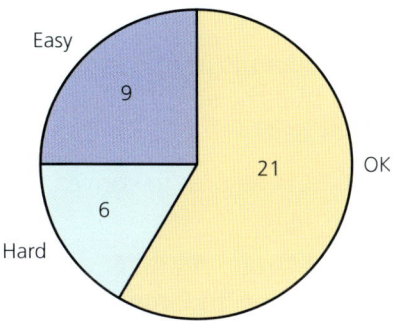

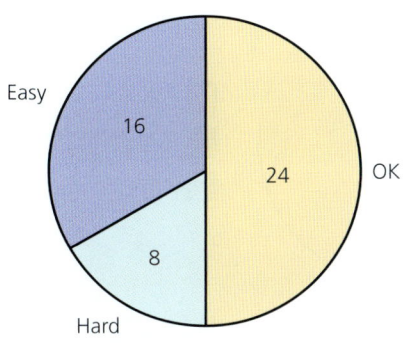

Comparing pie charts

You saw above that pie charts are an effective way of displaying data. Given a set of data displayed as a pie chart, it is important to be able make deductions from it.

For example, this pie chart shows the results of a survey of 36 students. They were asked whether they find mathematics easy, hard or OK.

From the pie chart we can see that most of these students find mathematics OK while the smallest number find it hard.

Another group of students were asked the same question. This pie chart shows their responses.

Looking at the two pie charts, it is easy to see that a smaller fraction of this group of students find mathematics OK than in the first group. But it is not so easy to compare the fractions who find it hard. There are more students (48) in the second group than in the first group (36) so we cannot just compare the numbers on the sectors. To be able to compare the data in the two pie charts, we need to use a common scale. For example, we can use percentages. The percentages for the first group are worked out in this table.

Group 1

	Frequency	Fraction of total	Percentage
Easy	9	$\frac{9}{36}$	$\frac{9}{36} \times 100 = 25\%$
OK	21	$\frac{21}{36}$	$\frac{21}{36} \times 100 = 58\%$
Hard	6	$\frac{6}{36}$	$\frac{6}{36} \times 100 = 17\%$

We can see that 17% of the students in this group find mathematics hard.

The percentage table produced from the pie chart for the second group is shown on the next page.

SECTION 1

Group 2

	Frequency	Fraction of total	Percentage
Easy	16	$\frac{16}{48}$	$\frac{16}{48} \times 100 = 33\%$
OK	24	$\frac{24}{48}$	$\frac{24}{48} \times 100 = 50\%$
Hard	8	$\frac{8}{48}$	$\frac{8}{48} \times 100 = 17\%$

From this table we can see that 17% of the students in this group also find mathematics hard. This shows us that the same percentage of students in each group find mathematics hard.

To compare pie charts with different totals, it is helpful to convert the frequency of each category into a percentage.

These pie charts show the favourite sports of Group A and Group B. There are 36 students in Group A, and 30 students in Group B.

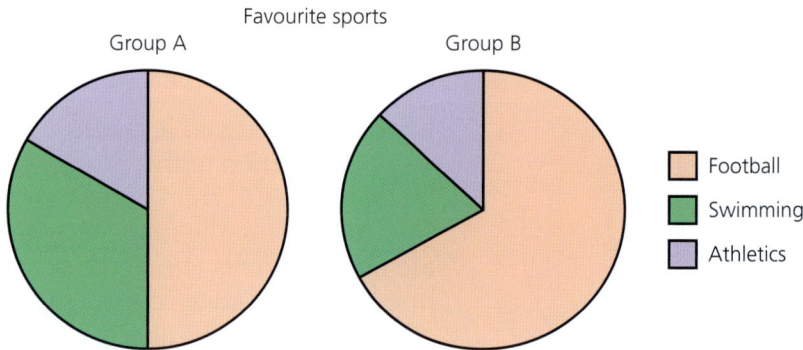

> Remember that there are 360° in a full circle.

Looking at the pie charts you can see that football is the most popular sport in both groups. But to find out how many students prefer football you will need to measure the angle and work out the fraction of each compared with the full circle.

The frequencies for Group A are worked out in this table.

Group A

	Angle	Fraction of total	Frequency
Football	180°	$\frac{180}{360}$	$\frac{180}{360} \times 36 = 18$
Swimming	120°	$\frac{120}{360}$	$\frac{120}{360} \times 36 = 12$
Athletics	60°	$\frac{60}{360}$	$\frac{60}{360} \times 36 = 6$

We can see that 18 students in Group A preferred football.

Here is the table for the second group.

Group B

	Angle	Fraction of total	Frequency
Football	241°	$\frac{241}{360}$	$\frac{241}{360} \times 30 = 20$
Swimming	72°	$\frac{72}{360}$	$\frac{72}{360} \times 30 = 6$
Athletics	47°	$\frac{47}{360}$	$\frac{47}{360} \times 30 = 4$

From this table we can see that 20 students in Group B preferred football.

Exercise 7.3

1. The hair colours of two groups of students are shown in the pie charts below.

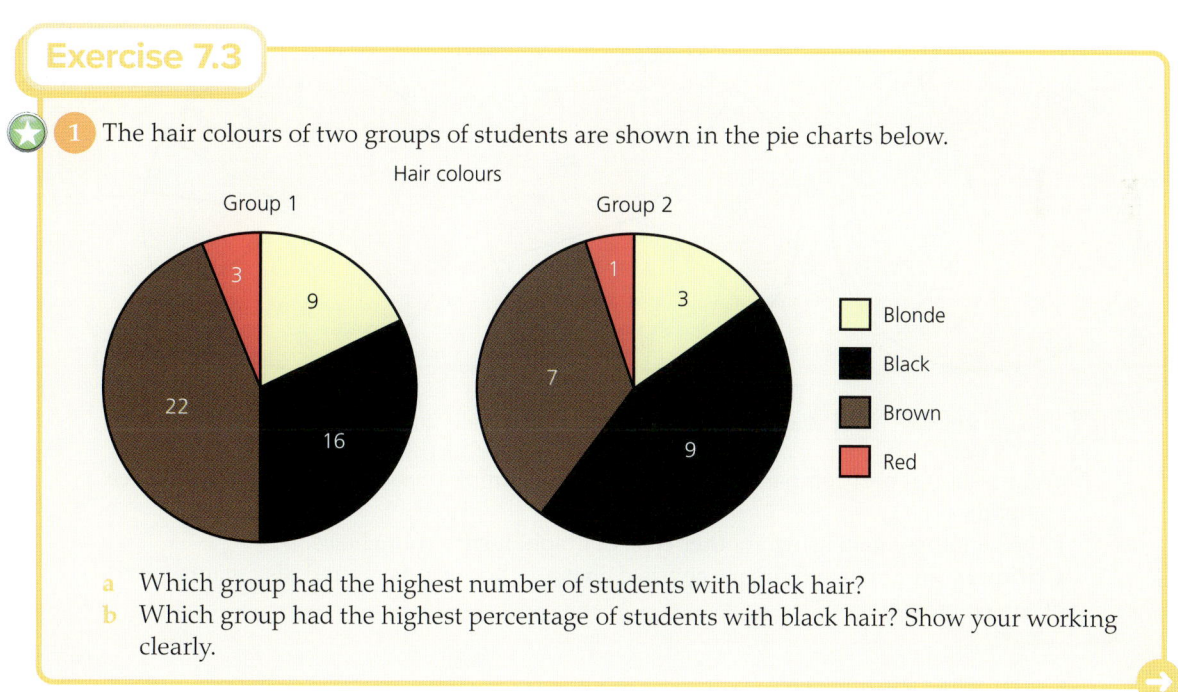

a Which group had the highest number of students with black hair?
b Which group had the highest percentage of students with black hair? Show your working clearly.

SECTION 1

   c A student in Group 1 looks at the pie charts and says, 'I think Group 1 has a greater proportion of students with blonde hair than Group 2'.
Find out whether the student is correct. Justify your answer.

2 Students in two different schools (X and Y) sit the same mathematics exam. The numbers of students achieving each grade (A to E) are shown in the pie charts below.

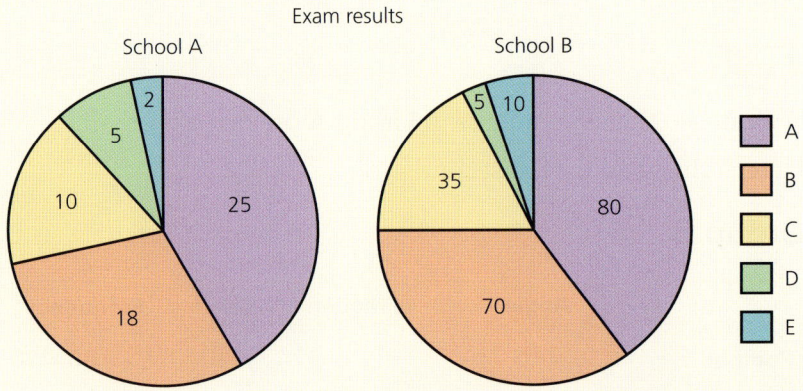

   a Which school achieved the highest percentage of A grades?
   b Which school achieved the highest percentage of B grades?
   c Write a brief comparison between the results of the two schools.

3 Three companies (X, Y and Z) publish data about the numbers of men and women they employ and their ages. These pie charts show a summary of the data.

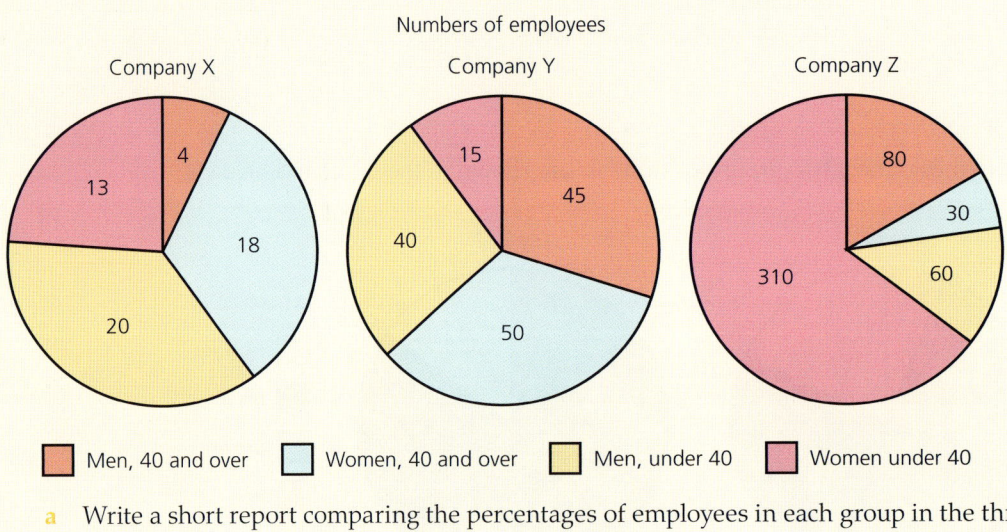

   a Write a short report comparing the percentages of employees in each group in the three companies.
   b Represent these data using another type of graph. Justify your choice.
   c **Critique** your graph and the pie charts. What are the advantages and disadvantages of each representation?

7 Recording, organising and representing data

Stem-and-leaf diagrams

You will already be familiar with the fact that grouped bar charts can be used for discrete data. The problem with this type of grouped data is that you are not able to see the individual values within the group and so lose some of the data's detail. A **stem-and-leaf diagram** is a special type of grouped bar chart, which has the advantage that the individual data values can still be seen in the diagram.

Worked example

These data record the ages of people on a bus to the seaside.

2	4	25	31	3	23	24	26	37	42
60	76	33	24	25	18	20	77	5	13
18	13	15	49	70	48	27	25	24	29

Display the data on a stem-and-leaf diagram.

```
0 | 2 3 4 5
1 | 3 3 5 8 8
2 | 0 3 4 4 4 5 5 5 6 7 9
3 | 1 3 7
4 | 2 8 9
5 |
6 | 0
7 | 0 6 7
```

The diagram must have a key to explain what the stem means. If the data were 1.8, 2.7, 3.2 etc., the key would state that '2 | 7 means 2.7'.

Key

3 | 1 means 31 years

The diagram must have a key to explain what the stem means. If the data were 1.8, 2.7, 3.2 etc., the key would state that '2 | 7 means 2.7'.

The shape of a stem-and-leaf diagram makes it quite straightforward to compare two sets of results side by side.

61

SECTION 1

Exercise 7.4

1. Twenty students sit two mathematics tests A and B. Each test is marked out of 50. The results are shown below.

Test A									
8	10	12	16	20	21	21	23	24	26
29	34	34	35	36	36	38	38	41	42

Test B									
18	26	27	27	29	34	34	35	35	36
38	38	39	41	41	43	45	48	48	49

   a Display the data on two separate stem-and-leaf diagrams.
   b One pupil thinks that Test A was harder than Test B. Comment on this by comparing the two diagrams.

2. A basketball team plays 25 matches during one season. They keep a record of the number of points they score in each game and the number of points scored against them. The results are shown below.

Points for					Points against				
64	72	84	46	53	51	72	40	66	69
69	62	71	71	79	43	58	81	78	60
80	47	53	69	69	60	42	57	69	74
56	82	84	78	78	40	41	72	66	54
72	68	66	54	64	72	66	51	53	67

LET'S TALK
Which other types of graph could be used for these data?

   a Draw two stem-and-leaf diagrams showing the numbers of points for and against the team during the season.
   b By looking at the shapes of the diagrams, is the team likely to have won more games than they lost, or the other way round? Justify your answer.

7 Recording, organising and representing data

 3 Fifteen people take part in a fitness assessment. Their pulse rates are taken before and after exercise. The results are recorded in these stem-and-leaf diagrams.

5						5	1	3	4	
5	8					5	6	9		
6	1	3				6	0	3	3	4
6	8	8	9			6	7	8	8	
7	0	2	3	4		7	1			
7	7					7	9	9		
8	1	3				8				
8	9					8				
9	3					9				

Key
6 | 8 means 68 beats per minute

a Calculate the mean, median, mode and range for the pulse rates for both diagrams.
b Which diagram is likely to show the readings taken after exercise? Justify your answer using your answers in part (a) to support your reason.

Scatter graphs

Scatter graphs are particularly useful if we wish to see if there is a relationship between two sets of data. The two sets of data are collected. Each pair of values then form the coordinates of a point, which is plotted on a graph.

There are several types of relationship, depending on the arrangement of the points plotted on the scatter graph.

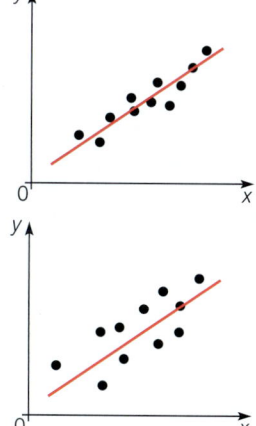

The points lie very close to the line of best fit.

As x increases, so does y.

Although there is direction to the way the points are lying, they are not tightly packed around the line of best fit.

As x increases, y tends to increase too.

SECTION 1

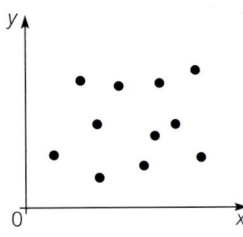

No relationship. As there is no pattern in the way the points are lying, there is no relationship between the variables x and y.

As a result, there can be no line of best fit.

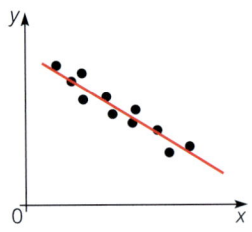

The points lie close to the line of best fit.

As x increases, y decreases.

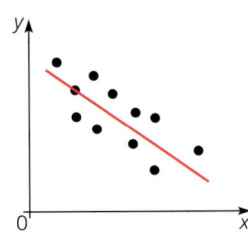

There is direction to the way the points are lying but they are not tightly packed around the line of best fit.

As x increases, y tends to decrease.

Worked example

Different cars have different engine sizes. A student decides to see if a car's engine size affects its fuel economy, i.e. how many kilometres it can travel (on average) on 1 litre of fuel. She randomly selects 14 cars and records their engine size (in litres) and their average fuel economy (in kilometres per litre). The results are given in the table below.

Engine size (litres)	3.2	1.8	1.8	2.4	1.5	2.0	1.9	2.8	1.8	2.0	2.0	4.3	1.9	2.8
Fuel economy (km/l)	10.2	14.2	13.6	14.0	15.7	11.9	16.2	12.3	15.7	12.5	14.5	9.8	16.1	11.1

7 Recording, organising and representing data

a Plot a scatter graph to represent the data.

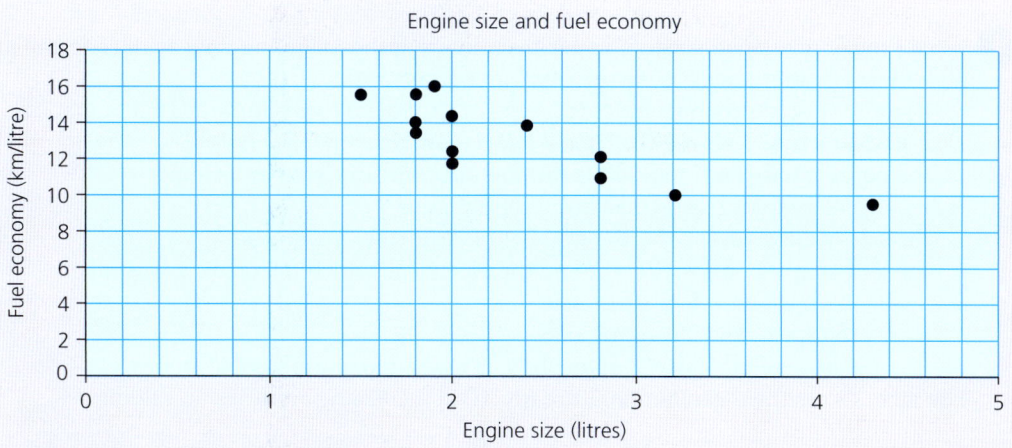

b Comment on any relationship that you see.

The points tend to lie in a diagonal direction from top left to bottom right. This suggests that as engine size increases then, in general, fuel economy decreases.

c If another car was found to have an engine size of 2.5 litres, approximately what fuel economy would you expect it to have?

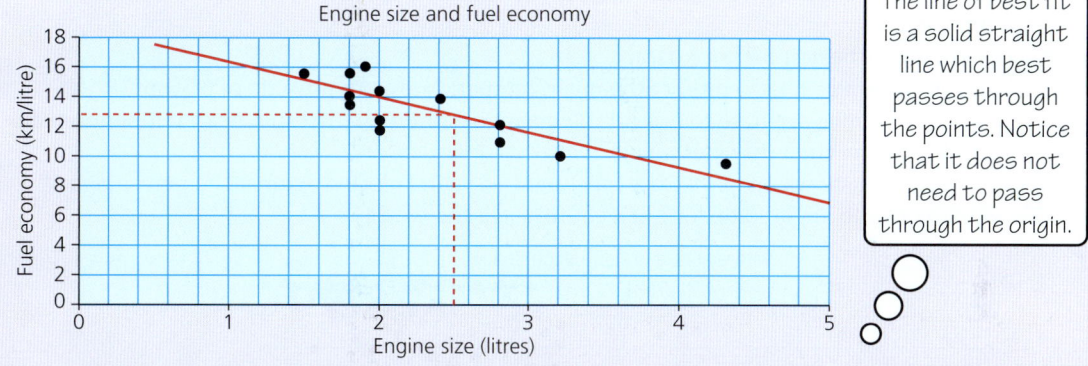

The line of best fit is a solid straight line which best passes through the points. Notice that it does not need to pass through the origin.

We need to assume that this car follows the trend set by the other 14 cars. To work out an approximate value for the fuel economy, we draw a line of best fit.

Then we draw a vertical line up from 2.5 litres on the horizontal axis until it reaches the line of best fit and read off the value for the fuel economy from the vertical axis.

The fuel economy of a car with an engine size of 2.5 litres is likely to be about 13 kilometres per litre.

SECTION 1

Exercise 7.5

1. What type of relationship might you expect, if any, if the following data were collected and plotted on a scatter graph? Give reasons for your answers.
   a a student's score in one mathematics exam and their score in another mathematics exam
   b a student's shoe size and the time it takes them to walk to school
   c the outdoor temperature and the number of ice creams sold by a shop
   d the age of a car and its second-hand selling price
   e the size of a restaurant and the number of customers it can seat
   f the temperature on a mountain and the altitude at which the reading was taken
   g a person's mass and their waist size
   h a car's top speed and its engine size.

2. The temperature of the Earth's atmosphere is measured at different altitudes.
   The table shows the mean temperature (in °C) at increasing altitudes (in kilometres).

Altitude (km)	0	1	2	3	4	5	6	7	8	9	10	11
Mean temperature (°C)	15	9	2	−5	−11	−17	−24	−31	−37	−43	−50	−56

   a Plot a scatter graph of the data with altitude on the x-axis and mean temperature on the y-axis.
   b Use the shape of your graph to describe the relationship (if any) between altitude and temperature.
   c Draw a line of best fit for your data.
   d Use your line of best fit to predict the mean temperature at an altitude of 6.5 km.
   e Why may it not be accurate to extend the line of best fit to predict the mean temperature at altitudes greater than the ones given?

3. This scatter graph shows people's life expectancy (in years) plotted against the percentage of the population who are literate, for most countries in the world.

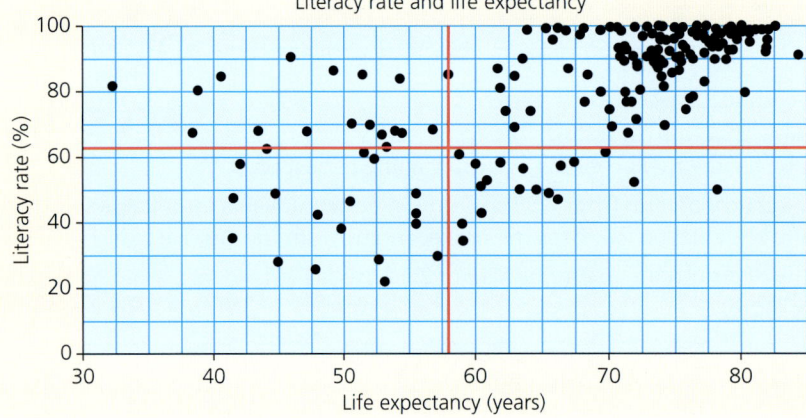

7 Recording, organising and representing data

a A horizontal and a vertical line have been drawn over the graph, splitting it into four sections. Describe the life expectancy and literacy levels in those countries found in the bottom left section.
b Describe the life expectancy and literacy levels in those countries found in the bottom right section.
c Use the graph to describe the relationship (if any) between life expectancy and literacy levels.

4 The mean temperature (in °C) and the mean rainfall (in millimetres) in Rio de Janeiro in Brazil are recorded over a 12-month period. The data are shown in the table.

Month	Jan	Feb	Mar	Apr	May	Jun	Jul	Aug	Sep	Oct	Nov	Dec
Mean temperature (°C)	26.2	26.5	26.0	24.5	23.0	21.5	21.3	21.8	21.8	22.8	24.2	25.2
Mean rainfall (mm)	114	105	103	137	86	80	56	51	87	88	96	169

a What are the highest and the lowest mean monthly temperatures?
b Using an appropriate scale on each axis, draw a scatter graph to show the relationship between mean temperature and mean monthly rainfall.
c Describe the relationship (if any) between the two variables.

The mean temperature (in °C) and the mean rainfall (in millimetres) are also recorded in Reykjavik in Iceland. The data are shown in the table.

Month	Jan	Feb	Mar	Apr	May	Jun	Jul	Aug	Sep	Oct	Nov	Dec
Mean temperature (°C)	−0.5	0.4	0.5	2.9	6.3	9	10.6	10.3	7.4	4.4	1.1	−0.2
Mean rainfall (mm)	76	72	82	58	44	50	52	62	67	86	73	79

d Draw a scatter graph showing the mean temperature and the mean monthly rainfall for Reykjavik.
e Is there a relationship between the two variables in Reykjavik?
f How does the data differ between the two locations?

In question 4 of Exercise 7.5 scatter graphs were plotted in order to see if there was a relationship between the mean monthly temperature and mean monthly rainfall in the two cities of Rio de Janeiro and Reykjavik.

As the months are also given, it is possible to plot a line graph to see how each variable changes over the course of the year. A line graph involving time is known as a **time series graph.**

SECTION 1

> **Worked example**
>
> The data below represent the mean monthly temperature (°C) for the two cities of Rio de Janeiro and Reykjavik.
>
> | Month | Jan | Feb | Mar | Apr | May | Jun | Jul | Aug | Sep | Oct | Nov | Dec |
> |---|---|---|---|---|---|---|---|---|---|---|---|---|
> | Mean temperature for Rio de Janeiro (°C) | 26.2 | 26.5 | 26.0 | 24.5 | 23.0 | 21.5 | 21.3 | 21.8 | 21.8 | 22.8 | 24.2 | 25.2 |
> | Mean temperature for Reykjavik (°C) | −0.5 | 0.4 | 0.5 | 2.9 | 6.3 | 9.0 | 10.6 | 10.3 | 7.4 | 4.4 | 1.1 | −0.2 |
>
> a On the same axes plot a time series graph for each city.
>
>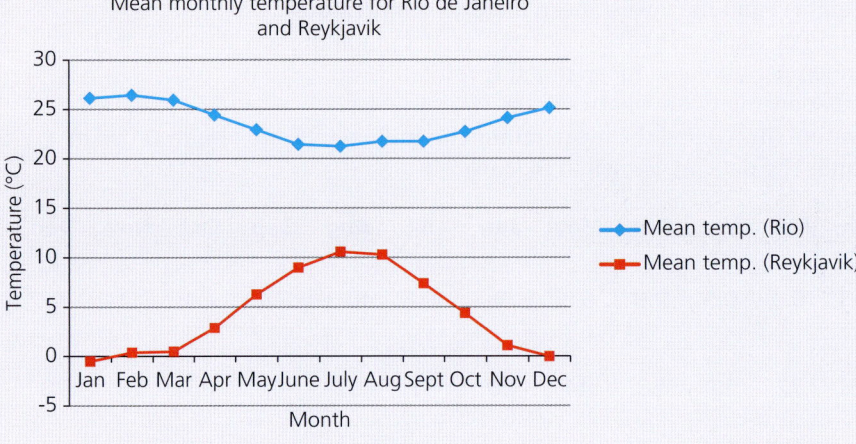
>
> b Which country has a more consistent temperature throughout the year?
>
> Rio de Janeiro as its time series graph does not change as much as Reykjavik's.

Time series graphs, as the name implies, are good for seeing changes over a period of time.

Decisions and interpretations

Drawing graphs and knowing which graphs are appropriate for the type of data is important. However, equally important is knowing how to interpret them. This section will focus on these aspects of data handling.

7 Recording, organising and representing data

> **Worked example**
>
> The wingspans (in centimetres) of male and female adult birds of a particular species are given in this table.
>
> | Wingspan (cm) | Frequency | |
> | --- | --- | --- |
> | | Male | Female |
> | $10 \leqslant w < 10.5$ | 3 | 8 |
> | $10.5 \leqslant w < 11$ | 4 | 10 |
> | $11 \leqslant w < 11.5$ | 10 | 8 |
> | $11.5 \leqslant w < 12$ | 9 | 4 |
> | $12 \leqslant w < 12.5$ | 4 | 0 |
>
> **a** Show both sets of data on a frequency graph.
>
>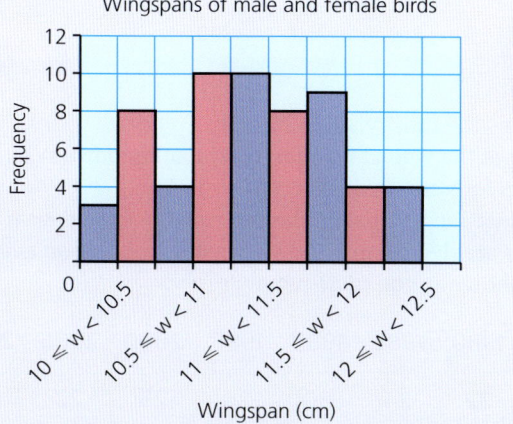
>
> **b** What is the modal wingspan for (i) the male birds, and (ii) the female birds?
>
> The modal group is the group with the highest frequency.
>
> **i)** The modal wingspan for the male birds is $11\,\text{cm} \leqslant w < 11.5\,\text{cm}$.
>
> **ii)** The modal wingspan for the female birds is $10.5\,\text{cm} \leqslant w < 11\,\text{cm}$.
>
> **c** Another bird of the same species is caught. Its wingspan is 11.8 cm. Is it possible to decide whether it is male or female from this information?
>
> It is not possible to decide for certain, as this wingspan is within both of the ranges. However, it is more likely to be male than female.

SECTION 1

Exercise 7.6

1. The bar chart below shows the numbers of people visiting a small museum each day during one week.

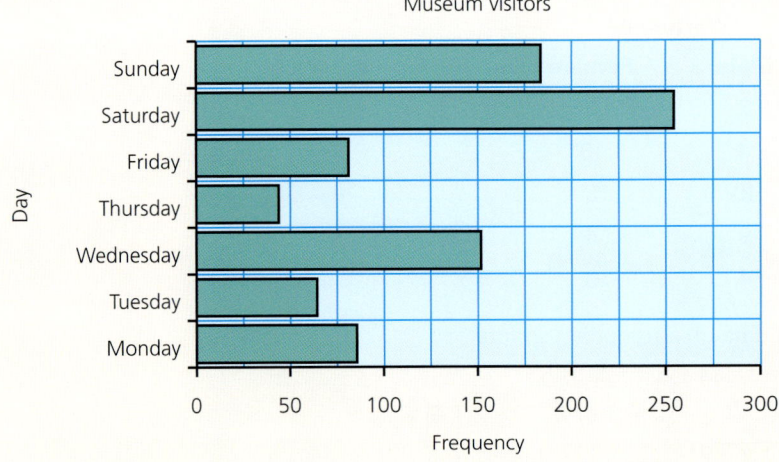

a. Give a possible reason why Saturday has the most visitors.
b. The museum is open for longer on one of the weekdays than on the other days. Which weekday is it likely to be? Give a reason for your answer.

2. A conservation group has studied the population of a particular mammal over a 50-year period. They are worried that the numbers are decreasing and survey the population every five years. The results of their survey are shown in the table below.

Year	1960	1965	1970	1975	1980	1985	1990	1995	2000	2005	2010
Population (1000s)	46	50	40	35	32	24	30	26	25	22	35

a. Draw a time series graph for the data.
b. By commenting on the shape of your graph, decide whether the conservationists' concern is valid or not.

7 Recording, organising and representing data

3. The numbers of rabbits and foxes in a national park are counted each year. Their numbers are shown in the line graphs below.

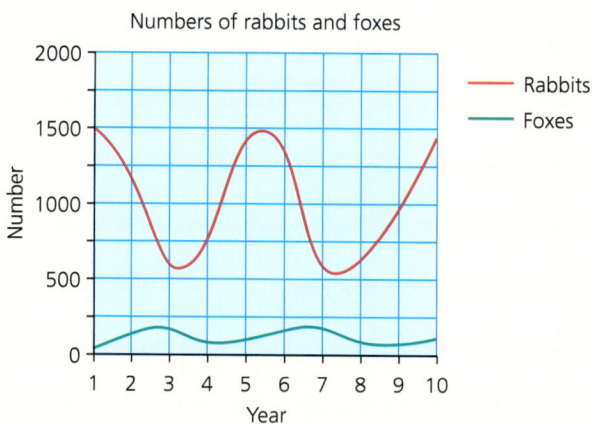

a   Describe the shape of the graph for the rabbits.
b   Compare the shapes of the two graphs.
c   Give a possible reason for the relationship between the shapes of the two graphs.

4. A company is going to install air conditioning in its offices. It will only be switched on when the outside temperature is higher than 28 °C. To estimate how much it will cost to operate, the outside temperature is recorded every four hours over a two-day period. The results are shown in the table below.

	Day 1						Day 2					
Time	00:00	04:00	08:00	12:00	16:00	20:00	00:00	04:00	08:00	12:00	16:00	20:00
Temperature (°C)	16	14	20	32	35	25	17	16	22	30	34	25

a   Draw a time series graph showing the temperature over the two days.
b   Use your graph to estimate the times on the first day when the outside temperature was 28 °C.
c   Estimate how long the temperature was higher than 28 °C on the first day.
d   Estimate how long the temperature was higher than 28 °C on the second day.

SECTION 1

5. Two classes take the same mathematics test, which is marked out of 10. One class is set by ability, and the other is a mixed ability class. These two graphs show the results for each of the classes.

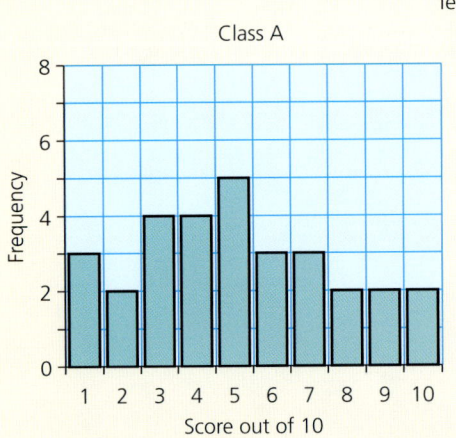

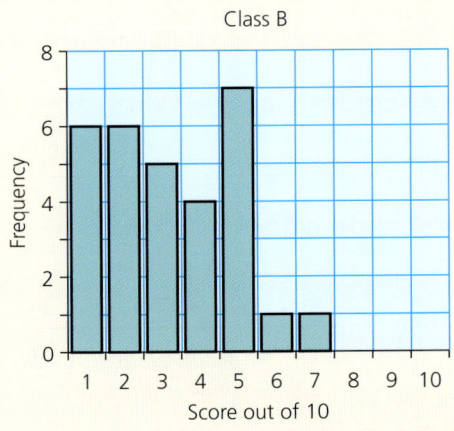

a   What is the modal score for each class?
b   What is the range of scores for each class?
c   Which class is likely to be the class which is set by ability? Give a **convincing** reason for your answer.

> A census is a survey that goes to every household in the country and asks the inhabitants lots of questions about them.

6. The population of the USA has been recorded through census data since 1830.
This table shows the population (in millions) at 10-year intervals.

Year	1830	1840	1850	1860	1870	1880	1890	1900	1910
Population (millions)	12.9	17.1	23.2	31.4	38.6	50.2	63.0	76.2	92.2
Year	1920	1930	1940	1950	1960	1970	1980	1990	2000
Population (millions)	106.0	123.2	132.1	151.3	179.3	203.3	226.5	248.7	281.4

a   What type of graph would be the most appropriate to display these data? Give a reason for your answer.
b   Plot the graph you chose in part (a).
c   Use your graph to estimate the population of the USA in 1975.
d   Use the shape of your graph to describe the change in the population of the USA since 1830.
e   Use your graph to deduce when the population of the USA was growing at its fastest. Justify your answer.

Now you have completed Unit 7, you may like to try the Unit 7 online knowledge test if you are using the Boost eBook.

8 Properties of three-dimensional shapes

- Understand and use Euler's formula to connect the number of vertices, faces and edges of 3D shapes.
- Use knowledge of area and volume to derive the formula for the volume of a triangular prism. Use the formula to calculate the volume of triangular prisms.
- Use knowledge of area, and properties of cubes, cuboids, triangular prisms and pyramids, to calculate their surface area.

Euler's formula

Leonard Euler (1707–1783) was a famous Swiss mathematician. He is often considered one of the greatest mathematicians, along with Sir Isaac Newton.

He studied many areas of mathematics, but also in Physics, Astronomy and Engineering.

This section looks at one of the beautiful formulae he discovered to do with the properties of **polyhedra**.

> Polyhedra means more than one polyhedron.

> Remember, a polygon is a 2D shape made up of straight edges. Examples include quadrilaterals, pentagons, hexagons etc.

A **polyhedron** is a three-dimensional shape, **characterised** by the fact that all its **faces** are flat. It is a type of **polygon**.

Different parts of a polyhedron have specific names as shown in the diagram of a cube below.

> Remember, more than one vertex are known as vertices.

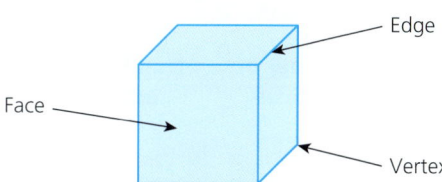

Worked example

Of the two shapes below, decide whether both, one or none are polyhedral. Justify your choice.

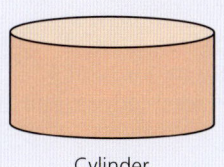
Cylinder

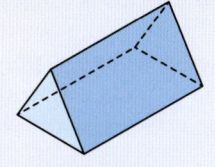
Triangular prism

The cylinder is not a polyhedron while the triangular prism is. This is because all the faces of a polyhedron must be a 2D polygon. The cylinder has two circular faces (not polygons) and a curved face.

SECTION 1

Exercise 8.1

1 The diagrams below show several 3D shapes and names.
   a Using the internet as a resource if necessary, match each of the shapes to their correct name.
   b Which of the shapes are **classified** as polyhedral?

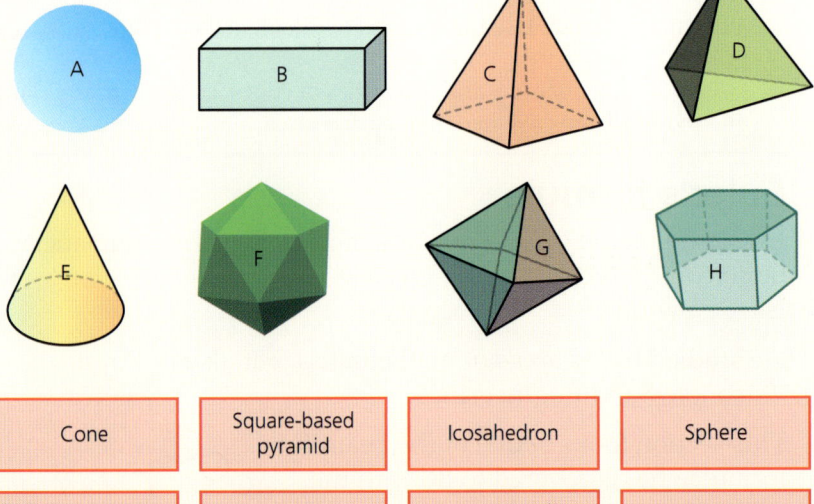

| Cone | Square-based pyramid | Icosahedron | Sphere |
| Cuboid | Tetrahedron | Hexagonal prism | Octahedron |

LET'S TALK
Many of these shapes belong to other families of shapes. Discuss what these might be.

2 a Draw a polyhedron not already drawn in questions 1 and 2 above.
   b What is the name of your polyhedron?

3 a For each of the polyhedra in questions 1–3, count the number of faces, edges and vertices and enter the results in a table similar to the one below. One is completed for you.

Name	Number of faces	Number of vertices	Number of edges
Triangular prism	5	6	9

   b Can you spot a **general** rule linking the number of faces, vertices and edges for each of the polyhedra in your table? If so, describe it in words.
   c Using F for the number of faces, V for the number of vertices and E for the number of edges express your rule described in part (b) as a formula using algebra.

4 a Using a book or the internet, find two more polyhedra. Write down their names.
   b Work out the numbers of faces, vertices and edges for your two polyhedra.
   c Check that the results you wrote for part b) fit the formula you worked out in question 4.

5 Explain why your rule will not work for a cylinder or a cone. Use the words 'faces', 'edges' and 'vertices' in your explanation.

8 Properties of three-dimensional shapes

Convex polyhedra

You found in the previous section that there is a relationship between the number of faces (*F*), vertices (*V*) and edges (*E*) of polyhedra. As a formula this can be written as:

$$V + F - E = 2$$

This is known as **Euler's formula** and is true for most types of polyhedron. It does not work, however, for other 3D shapes.

Exercise 8.2

1. Complete the following sentences:
   a. The corners of a polyhedron are called ....
   b. The flat side of a polyhedron is called a ....
   c. Two corners are joined by an ....
2. a. Find the number of edges in polyhedra with the following properties:
      i) six vertices and five faces
      ii) ten vertices and seven faces.
   b. Find the number of vertices in polyhedra with the following properties:
      i) ten faces and twenty-four edges
      ii) seven faces and twelve edges.
3. Sadiq is holding a polyhedron. He states that it has 8 vertices and 12 edges. How many faces does it have?
4. Arabella is holding a polyhedron. She says that it has 12 vertices, 7 faces and 15 edges.
   a. Explain why she must have miscounted at least one of the values.
   b. If the numbers of faces and edges are correct, how many vertices must the shape have?
5. Natalia claims that she has found a polyhedron (shown) that does not follow Euler's rule.
   She says that it has 0 vertices, 3 faces and 2 edges and using the formula $V + F - E = 2$ this means $0 + 3 - 2 = 1$.
   Explain her mistake.

However, Euler's formula does not work for every type of polyhedron either. All the polyhedra so far have been examples of **convex polyhedra**.

SECTION 1

For a polyhedron to be convex, a line joining any two vertices falls completely on or within the shape itself. Looking at the cuboids shown below, in the first case the line joining two vertices runs along the edge of the cuboid, while in the second example it passes through the cuboid.

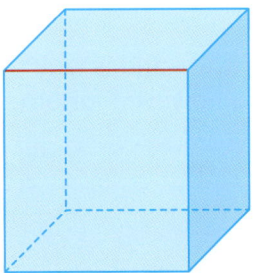 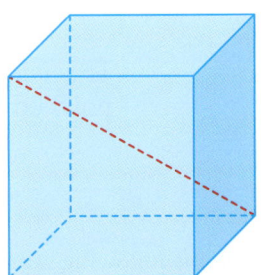

> **KEY INFORMATION**
> A pentagonal-based pyramid is simply a pyramid with its base in the shape of a pentagon.

This is not the case for all polyhedra.

The diagram below shows a **dodecahedron**, which is a convex polyhedron with 12 faces. Each face of the dodecahedron is a **regular pentagon**. If a **pentagonal-based pyramid** is added to each face, the shape produced is called a **stellated** dodecahedron.

The word stellated means star shaped.

Dodecahedron   Stellated dodecahedron

> **LET'S TALK**
> How can we show that a stellated dodecahedron will not fit Euler's formula without having to count all the faces, vertices and edges?

If a line is drawn between the two of the peaks of the stellated dodecahedron it will not pass through the shape itself.

More analysis of these types of polyhedra is beyond the syllabus for this book.

> **LET'S TALK**
> What other regular, non-convex polyhedra are there? Research this with a friend.

8 Properties of three-dimensional shapes

Triangular prisms

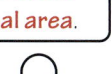

If a 3D shape is the same all the way through it is said to have a **constant cross-sectional area**.

You already know that a cuboid can look similar to this:

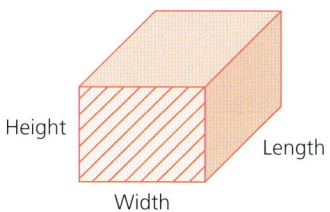

All faces are either squares or rectangles.

To calculate its volume, the length is multiplied by the width and height:

Volume of cuboid = Length × Width × Height

From the diagram we can see that Width × Height gives the area of the shaded end face.

Therefore, Volume of cuboid = Area of end face × Length

If the cuboid was sliced vertically, the shape of the end face would be the same all the way through.

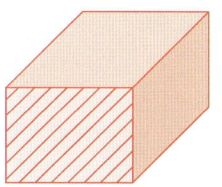

LET'S TALK
What other 3D shapes belong to the family of prisms? Can you sketch them?

This is known as the **cross-section** of the cuboid. A 3D shape with a constant cross-sectional area is called a **prism**. A cuboid therefore is a type of prism.

A type of prism you will have seen before is a **triangular prism**.

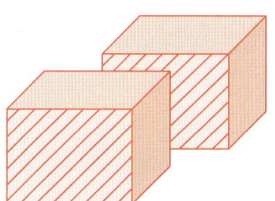

The cross-section is a triangle.

Volume of triangular prism = Area of cross-section × Length

$$= \frac{1}{2} \times \text{Base length} \times \text{Height} \times \text{Length}$$

SECTION 1

Worked example

a The triangular prism opposite has dimensions as shown.
Calculate its volume.
Volume of triangular prism = Area of cross-section × Length.
The area of the cross-section is the area of the triangular end face:

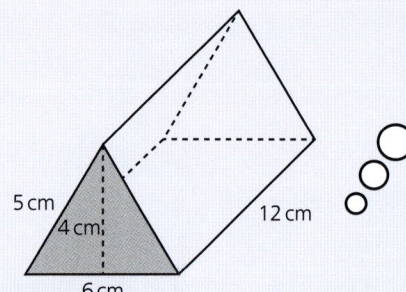

Area of cross-section = $\frac{1}{2}$ × Base length × Height

$= \frac{1}{2} \times 6 \times 4 = 12 \text{ cm}^2$

Therefore, volume is 12 × 12 = 144 cm³

> Note that to calculate the area of a triangle the perpendicular height is needed and not the length of the sloping edge. The 5 cm length is therefore not needed.

b The volume of a triangular prism opposite is 270 cm³.
Calculate its height.
Volume = Area of cross-section × Length
270 = Area of cross-section × 12
Therefore, rearranging the formula gives:
Area of cross-section = $\frac{270}{12}$ = 22.5 cm²

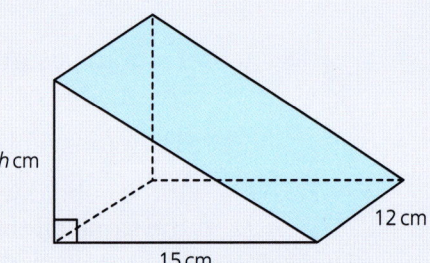

Area of cross-section = $\frac{1}{2}$ × Base length × Height

$22.5 = \frac{1}{2} \times 15 \times h$

$45 = 15 \times h$ (multiplying both sides of the formula by 2)

$\frac{45}{15} = h$ (dividing both sides by 15)

$h = 3$ cm

> The formula needs to be rearranged to make the height, h, the subject.

Exercise 8.3

1 Calculate the volumes of the following triangular prisms:

a
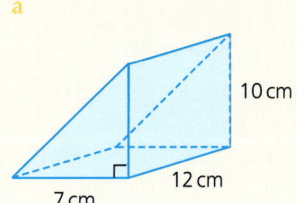

b
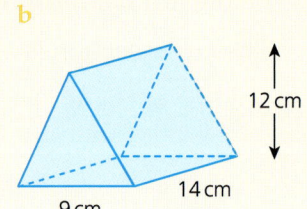

78

2 Calculate the missing quantities in each of the following triangular prisms.

	Base length (cm)	Height (cm)	Length (cm)	Cross-sectional area (cm²)	Volume (cm³)
a	10	6	8		
b	4	15	6		
c		12	6	48	
d	10	7			700
e		10		25	125

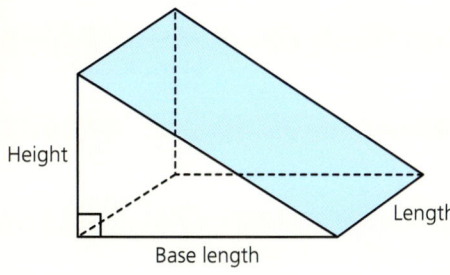

3 a Calculate the volume of the whole triangular prism shown on the right.
A horizontal cut is made half-way down the height of the prism as shown by the shaded slice. It divides the original prism into two pieces, A and B. Each length of prism A is half that of the original whole prism.
b What is the ratio of the volume of the two pieces, A : B? Give your answer in its simplest form.

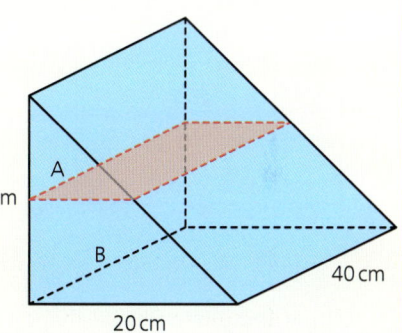

4 The volume of the triangular prism shown is 560 cm³. The area of the triangular cross-section is 40 cm². Calculate the values of x and y.

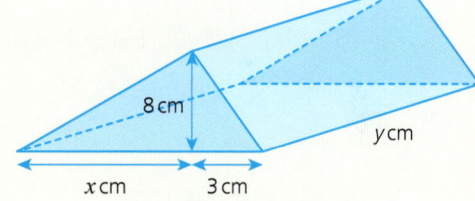

5 A water tank is in the shape of a cuboid. When the triangular prism shown is lowered into the water, the water level rises by 2 cm. Calculate the length of the prism.

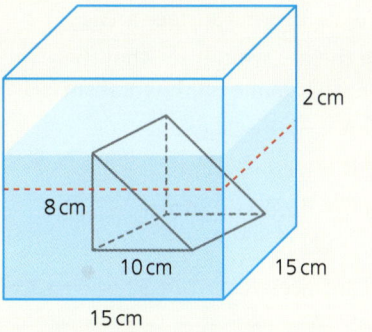

SECTION 1

Surface area

The surface area of a 3D shape is the total area of all of its faces. Therefore, knowledge of what types of face make up the 3D shape is important.

A net of an object is a 2D representation of that object. When a **net** is folded up, it makes that 3D object.

To do this it is sometimes helpful to visualise the net of the 3D shape.

Worked example

The net of a 3D object is shown.

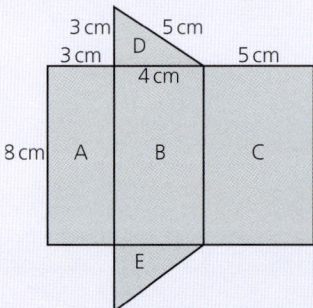

> **LET'S TALK**
> Six dimensions have been included in the diagram. Was it necessary to write all six to know the dimensions of the shape? Which ones are necessary?

a Draw and name the 3D shape.

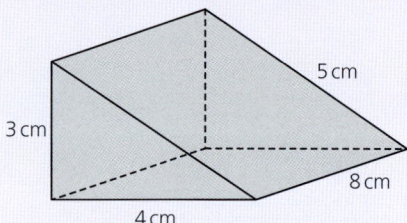

A triangular prism

b Calculate its total surface area.

From the net we can see that there are five faces, A, B, C, D and E, that make up the total surface area of the triangular prism.

Area A = 3 × 8 = 24 cm²
Area B = 4 × 8 = 32 cm²
Area C = 5 × 8 = 40 cm²
Areas D & E = $\frac{1}{2}$ × 4 × 3 × 2 = 12 cm²
Total surface area = 24 + 32 + 40 + 12 = 108 cm²

> As the triangles D & E are the same, the area of one of them is calculated and then the result just doubled.

8 Properties of three-dimensional shapes

Exercise 8.4

1 a Four nets are shown on the right. Some of them fold up to make a cube. Which one(s) do not make a cube? Justify your answer.
  b Draw another net for a cube different from the ones given opposite.
  c If the face of the cube has an edge length of 5 cm, calculate the total surface area of the cube.

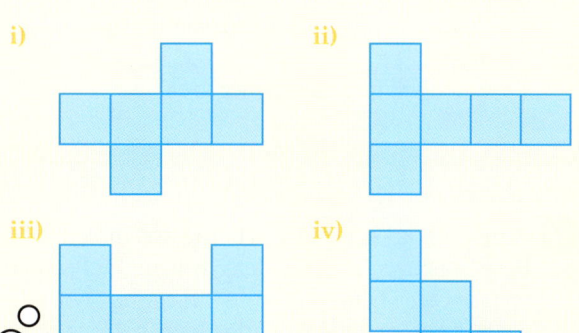

i) ii) iii) iv)

Cut out the nets from squared paper to check if necessary.

2 A cuboid is shown.
  a Draw two possible nets for the cuboid.
  b Calculate the total surface area of the cuboid.

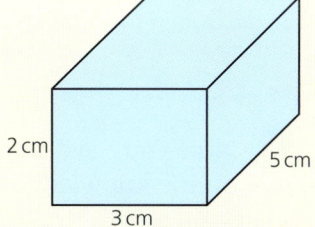

2 cm, 3 cm, 5 cm

3 A square-based pyramid is shown.
  a Calculate the area of one of its triangular faces.
  b Calculate the total surface area of the pyramid.

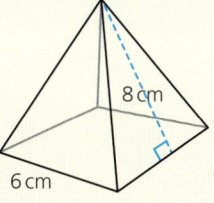

6 cm, 8 cm

4 The cuboid shown has a square cross-section and a height h cm.
  a If the total surface area is 594 cm², show that it can be given by the equation
  $594 = 162 + 36h$
  b Calculate the value of h.

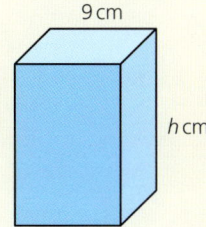

9 cm, h cm

SECTION 1

5 A triangular prism is shown opposite.
  a Draw and label a possible net for the triangular prism.
  b Calculate its total surface area.

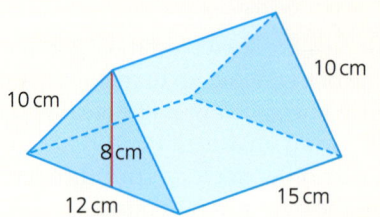

6 A modern art sculpture is being assembled. Its base is a cube A of edge length 10 m. Only five of its faces are visible as the one in contact with the ground cannot be seen. Another cube B of edge length 9 m is placed on top as shown.
  a Calculate the total area of the visible surfaces of the combined sculpture of A and B.
  b Successive cubes are placed on top of each other. The edge length of each decreases by 1 m each time. How many cubes is the final sculpture made from if the total area of the visible surfaces is 1420 m²?

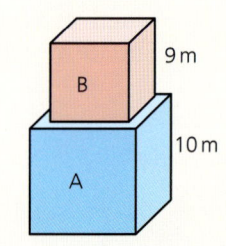

Now you have completed Unit 8, you may like to try the Unit 8 online knowledge test if you are using the Boost eBook.

9 Factors and multiples

- Understand factors, multiples, prime factors, highest common factors and lowest common multiples.
- Understand the hierarchy of natural numbers, integers and rational numbers.

Factors and multiples

You will remember from Stage 7 that **factors** of a number are all the whole numbers (positive integers) which divide exactly into that number. For example, the factors of 12 are 1, 2, 3, 4, 6 and 12 because 12 is divisible by all of them exactly.

LET'S TALK
Can you remember why 1 is not considered to be a prime number?

You will also know that a **prime number** is a number which has only two factors, these being 1 and the number itself. For example, 7 is a prime number as the only numbers that 7 is divisible by are 1 and 7.

Combining these two definitions we can see that of the factors of 12, some are also prime numbers:

    1   2   3   4   6   12

As 2 and 3 are factors of 12 and also prime numbers, they are known as the **prime factors** of 12.

Factor trees

One way of finding the prime factors of a number is to use a factor tree.

LET'S TALK
The number 45 in the factor tree was originally split as 9 × 5. But 45 could have been written as a product of two different numbers. Would this have affected the final answer?

Worked example

a Find the prime factors of 45.

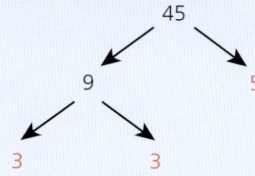

The prime factors of 45 are 3 and 5.

83

SECTION 1

> Note that each number is split into two numbers which multiply to give that number. For example, $45 = 9 \times 5$. A branch is not continued further when a prime number is reached.
>
> **b** Express 45 as a product of prime factors.
>
> From the factor tree we can see that 45 can be written as $3 \times 3 \times 5$ which, because they are all prime numbers, means we have written 45 as a product of its prime factors.
>
> This product can also be written using indices, i.e. $3 \times 3 \times 5 = 3^2 \times 5$.

The word 'product' means the answer to a multiplication.

Highest common factors and lowest common multiples

If we look at the number 24, we can work out that its factors are 1, 2, 3, 4, 6, 8, 12 and 24.

If we look at the number 28, we can work out that its factors are 1, 2, 4, 7, 14 and 28.

This information can be displayed in a Venn diagram:

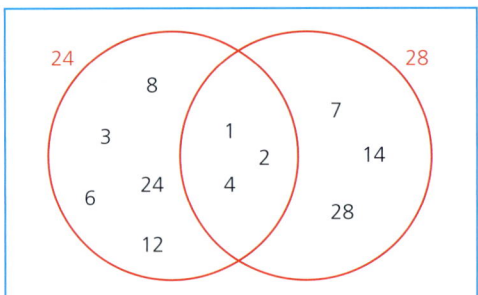

The overlap of the two circles shows those numbers which are factors of both 24 and 28. The number 4 is the largest of those. That is to say 4 is the **highest common factor (HCF)** of both 24 and 28.

The highest common factor of numbers can also be calculated using the results of factor trees.

9 Factors and multiples

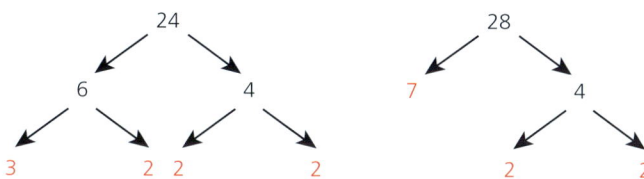

By looking at the prime factors of each number, we can see that the ones that are common to both are 2×2. Therefore, the highest common factor is $2 \times 2 = 4$.

The **lowest common multiple (LCM)** of two numbers refers to the smallest number which is in the times table of both numbers.

Looking at the two numbers 24 and 28.

The 24 times table can be written as 24, 48, 72, 96, 120, 144, **168**, 192, 216 …

The 28 times table can be written as 28, 56, 84, 112, 140, **168**, 196, 224 …

168 is the smallest number to appear in both times tables. So, 168 is therefore the lowest common multiple of 24 and 28.

This too can be calculated using the factor trees by using the greatest number of times each prime number appears and multiplying them together. For example:

The prime number 2 appears the most times as $2 \times 2 \times 2$ in the factors of 24.

The prime number 3 appears only once as a factor of 24.

The prime number 7 appears only once as a factor of 28.

The lowest common multiple is therefore $2 \times 2 \times 2 \times 3 \times 7 = 168$.

> **LET'S TALK**
> By using the difference of 24 and 28, is there another way of finding the lowest common multiple of those two numbers? Does your method work all the time?

SECTION 1

Exercise 9.1

1 a Find the factors of the following numbers:
   i) 50 ii) 72
   b What is the highest common factor of the two numbers?
   c Peter looks at his answers to part (a) and says, 'The bigger the number, the more factors it will have.' Is Peter correct? Give a **convincing** reason for your answer.

2 a i) Copy and complete the factor tree opposite.
   ii) Write 60 as a product of prime factors using indices.
   b i) Draw a different factor tree from the one above to find the prime factors of 60.
   ii) Compare your answers to parts (a) (ii) and (b) (i).
   c i) Draw another different factor tree to find the prime factors of 60.
   ii) Compare your answers to parts (a) (ii), (b) (i) and (c) (i). What conclusion can you make?

3 a Using a factor tree, find the highest common factor of the following pairs of numbers:
   i) 35 and 60 ii) 80 and 144 iii) 36 and 60
   b Find the lowest common multiple of each of the pairs of numbers in part (a).

4 Copy and complete the following factor trees:
   a (tree with 84 at top, branching down with 2, 2 at bottom)
   b (tree with 600 at top, 50 on left branch)

9 Factors and multiples

c i)

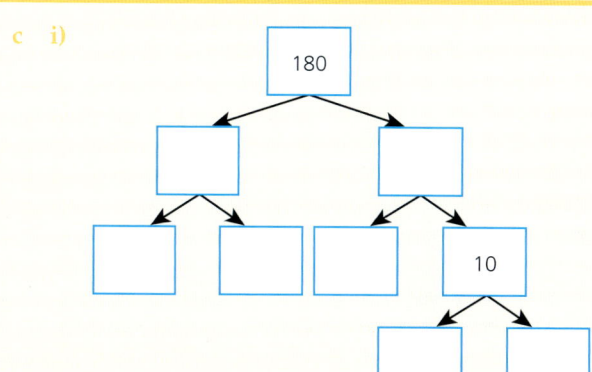

ii) Is there another possible arrangement of numbers for the given factor tree in part (i)? Justify your answer.

5 Two numbers, x and y, have a highest common factor of 7.
   a Give two possible values for x and y.
   b Give another possible value for x and y.
   c If the lowest common multiple of x and y is 210, calculate their values. Show your method clearly.

LET'S TALK
What other types of number group do you know?

Hierarchy of numbers

You will already be familiar with some groups of number such as even and odd numbers, and positive and negative numbers too. In this section you will be introduced to some other number groups.

An **integer** is a whole number such as 8, 44 and 426. But they can also be negative such as −12 or −56. Integers can therefore be subdivided into **positive integers** and **negative integers**.

Whether 0 is included as a natural number is still debated in the mathematical world. Some argue that it is a natural number, others say it isn't.

Zero is also **classified** as an integer and it is the only one that is neither positive nor negative.

A **natural number** is an integer greater than zero. They are often known as the counting numbers as they are 1, 2, 3, 4 etc.

LET'S TALK
Why are integers also rational numbers?

A **rational number** is any number that can be written as a fraction. Virtually all the numbers you will have encountered up until now are rational numbers. Using a calculator, if necessary, you can check, for example, that 0.3 is equivalent to the fraction $\frac{3}{10}$, or that $4.6251 = 4\frac{6251}{10000}$.

SECTION 1

> A terminating decimal is one which ends after a certain number of decimal places, e.g. 0.35 or 3.8267. A recurring decimal is one which repeats itself, e.g. 0.333333 etc.

In fact any **terminating decimal** or **recurring decimal** can be written as a fraction, but this will be covered in more detail in Stage 9.

Worked example

The table below shows six numbers and three different **classifications** of number.

	−8.4	−7	0	5	3.6	$\frac{5}{8}$	π
Natural number							
Integer							
Rational number							

Copy and complete the table by ticking which group each of the numbers belong to each group.

> π, as it is a decimal which never repeats itself, cannot be written as a fraction, so is not a rational number.
>
> 0 can also be considered a natural number by some mathematicians.

	−8.4	−7	0	5	3.6	$\frac{5}{8}$	π
Natural number				✓			
Integer		✓	✓	✓			
Rational number	✓	✓	✓	✓	✓	✓	

From the worked example you can see that if a number is a natural number it is also an integer and a rational number. The opposite is not true, however; not every integer or rational number is a natural number. Similarly, every integer is also a rational number but not necessarily vice versa.

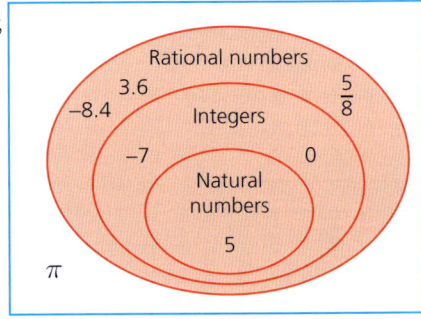

There is therefore a hierarchy of numbers. The results are shown in the Venn diagram.

9 Factors and multiples

Exercise 9.2

1. Which of the following statements are true?
   a. 8 is a natural number
   b. 6.5 is a natural number
   c. −12 is an integer
   d. 0 is an integer
   e. 19.1 is an integer
   f. 19.1 is a rational number
   g. $-\dfrac{13}{2}$ is a rational number

2. a. Which of the following statements are true?
   i) 15 is both a natural number and an integer
   ii) −100 is both an integer and a rational number
   iii) −56 is both a natural number and a rational number
   iv) 14.8 is a rational number but not a natural number
   b. Justify which of the statements above is/are false.

3. Which of the following statements are true?
   a. A rational number is always an integer.
   b. An integer is always a natural number.
   c. A natural number is always both an integer and a rational number.

4. Eight cards are arranged in ascending order of size. Two of the cards, A and B, are covered as shown:

   | −15 | −2.1 | A | 2 | $\dfrac{9}{4}$ | B | 8 | 9.1 |

   > Ascending order means numbers increasing in size from left to right.

   a. What card could A be if:
   i) it is an integer?
   ii) it is a natural number?
   b. What card could B be if:
   i) it is a natural number?
   ii) it is a prime number?
   iii) it is a rational number greater than 7?

5. Five cards, all with rational numbers on, are turned over. What numbers could be on the cards if they satisfy all of the conditions below?
   There are:
   - three integers
   - two natural numbers
   - two negative numbers
   - three positive numbers
   - one even number
   - two odd numbers
   - no prime numbers.

LET'S TALK
Can an even number be negative?

▶ Now you have completed Unit 9, you may like to try the Unit 9 online knowledge test if you are using the Boost eBook.

10 Complementary events

- Understand that complementary events are two events that have a total probability of 1.

Complementary events

The spinner below is unbiased, therefore landing on each section is equally likely.

As there are six sections, the probability of landing on each is $\frac{1}{6}$.

However, as there is a bigger fraction of one colour than others, the probability of getting each colour is not equally likely.

P(R) in this case means 'The probability of getting a red'. Similarly, P(Y) means 'The probability of getting a yellow'.

*A probability of 1 means that the **event** is certain. A probability of 0 means the event is impossible.*

The probability of the spinner landing on red is $\frac{3}{6} = \frac{1}{2}$

This can be written as $P(R) = \frac{1}{2}$

Similarly, $P(Y)$ is $\frac{2}{6} = \frac{1}{3}$ and $P(B)$ is $= \frac{1}{6}$

Adding together the probability of getting each colour gives $\frac{3}{6} + \frac{2}{6} + \frac{1}{6} = \frac{6}{6} = 1$

The **possible outcomes** are red, yellow and blue.

The sum of all the possible outcomes is always equal to 1.

In the spinner above we saw that $P(B) = \frac{1}{6}$; therefore, the probability of not getting blue must be $\frac{5}{6}$.

This can be written as $P(B') = \frac{5}{6}$

In **general**, $P(A') = 1 - P(A)$.

The notation P(B') means the probability of not getting B.

10 Complementary events

Worked example

The Venn diagram shows two groups of numbers, even numbers (E) and prime numbers (P).

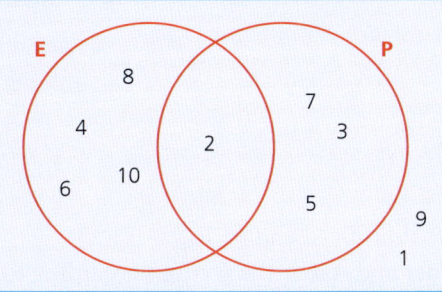

i) A number is picked at random, calculate P(P).

P(P) = $\frac{4}{10}$ as there are four prime numbers out of a possible ten numbers to choose from.

The probability can be written as $\frac{4}{10}$ or simplified to $\frac{2}{5}$.

ii) A number is picked at random, calculate P(E').

P(E') = $\frac{5}{10}$

This is calculated either by counting the numbers not in the 'even' circle or by

P(E') = 1 − P(E)

= 1 − $\frac{5}{10}$

= $\frac{5}{10}$

Exercise 10.1

1. A child's box of building blocks contains 15 red blocks (R) and 10 green blocks (G). A block is picked at random.
   a Calculate the probability of picking the following:
      i) P(R)
      ii) P(G)
      iii) P(G')
      iv) P(R')
   b Comment on your answers to parts (i) and (iii) and also on your answers to parts (ii) and (iv). Give a reasoned explanation for your observation.

2. If P(A) = $\frac{3}{10}$
   a Calculate P(A').
   b Give a **convincing** reason for your answer.

3. A temporary traffic light has two colours, red (R) and green (G). A driver knows that at any given time, the probability of getting a red light is P(R) = $\frac{1}{3}$.
   Calculate:
   a i) P(G)
   b ii) P(R')
   b What do you notice about your answers in part a? Justify your answer.

SECTION 1

 4 A hexagonal spinner is divided into six equilateral triangles as shown.

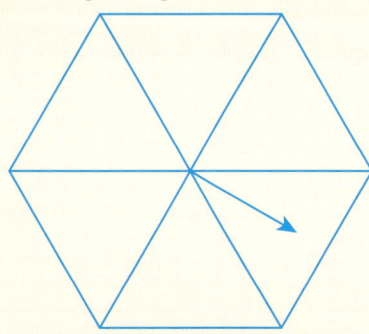

For each of the following questions, draw a new spinner and colour the sections according to the probabilities given.

a P(Green) = $\frac{1}{3}$

b P(Blue′) = $\frac{1}{2}$

c P(Red′) = 1

d P(Yellow′) = 0

 5 The Venn diagram shows a hierarchy of numbers. It shows natural numbers (N), integers (I) and rational numbers (R).

a Write one more number in each of the three rings of the Venn diagram.

The hierarchy of numbers was covered in Unit 9 of this book.

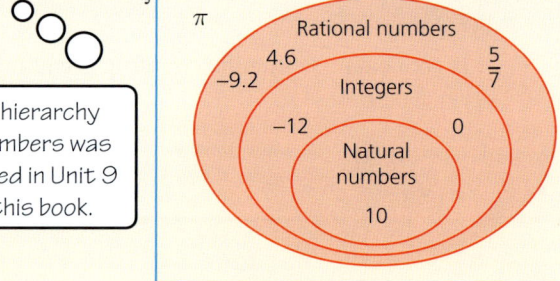

Use all ten numbers now in the Venn diagram:

b One number is picked at random; calculate the probability of picking an integer, P(I).

c Calculate the probability P(R′).

 Now you have completed Unit 10, you may like to try the Unit 10 online knowledge test if you are using the Boost eBook.

Section 1 – Review

1. Without using a calculator, and explaining your method, work out:
   a. 80×4
   b. 82×4
   c. $-82 \times (-4)$

2. a. Name all the quadrilaterals which have two pairs of parallel sides.
   b. Which of your quadrilaterals is at the top of the hierarchy? Justify your answer.

3. The height (cm) of 20 students are given in the table below. Their gender (B or G) is also included in the data.

1	2	3	4	5	6	7	8	9	10
G	B	B	B	G	B	G	G	B	G
154	161	158	172	148	163	164	165	158	147
11	12	13	14	15	16	17	18	19	20
B	G	G	B	B	G	B	G	G	B
159	161	153	161	163	157	155	149	171	170

   a. A sample of every third student is taken. These have been highlighted.
      Explain whether this method of sampling produces a representative sample of the population.
   b. Describe another type of sampling which could be used. Justify your choice.

4. a. Calculate the area of the trapezium.

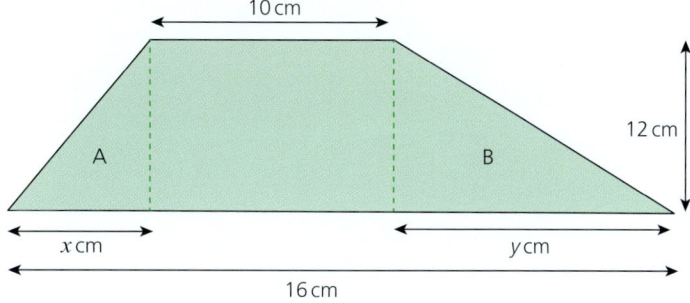

   b. If $y = 2x$ calculate the area of triangles A and B.

SECTION 1

5 Use a scientific calculator to work out
   a $\frac{(9-4)^2}{10} - \frac{1}{2}$
   b $5 + \frac{\sqrt[3]{270-54}}{2}$

6 The formula for the circumference of a circle is $C = 2\pi r$.
   a Which are the variables in the formula?
   b Which are the constants?
   c Calculate the radius of a circle with a circumference of 200 cm. Give your answer correct to 1 decimal place.

7 A pizza van keeps a record of the number of each type of pizza it sells during a week. These are shown below.
   - Margherita   57
   - Pepperoni   21
   - La Reine   105
   - Fiorentina   81
   - Giardiniera   36

   a Calculate the percentage of each pizza sold then convert the percentage to degrees of a circle.
   b Draw a pie chart of the data.

8 Euler's formula for polyhedra states that: $V + F - E = 2$
   a Explain what each variable in the formula represents.
   b Show that the formula works for the triangular prism opposite.

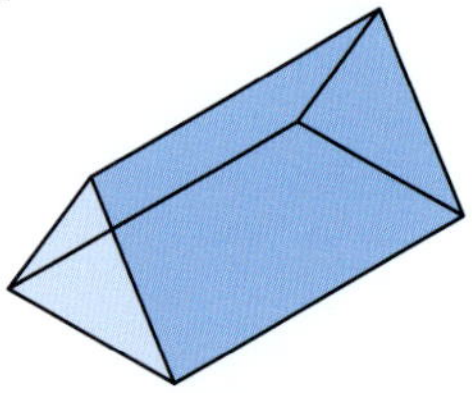

9 Using a factor tree find for the numbers 48 and 70
   a their highest common factor
   b their lowest common multiple.

10 A biased six-sided dice has a probability of getting a six of $\frac{5}{8}$. Calculate P(6′).

SECTION 2

History of mathematics – The Indians

'Mathematics knows no races or geographic boundaries. It is a game played within certain rules with meaningless marks on paper.'

David Hilbert

An unknown genius invented numbers. These strange shapes soon replaced tally marks. Numbers meant that arithmetic could be invented and then played with. Our present number system originated in India with the Brahim numerals about 300 BCE.

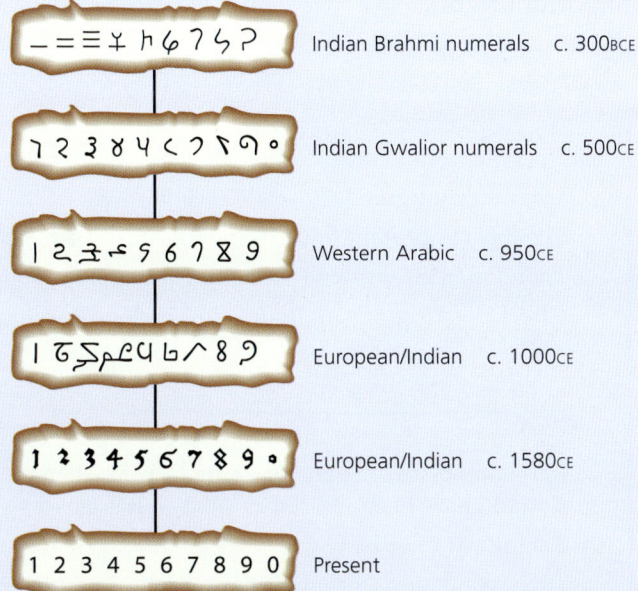

Mathematics from the Indian Subcontinent (India/Pakistan/Bangladesh) dates from 1300 BCE. In 1300 BCE a teacher named Laghada used geometry to help with architectural plans for cities and trigonometry for his astronomical calculations. He also solved complex algebraic equations.

Much later (about 600 CE) another teacher, Aryabhata, worked on approximations for pi and worked on the trigonometry of the sphere. He worked out that not only did the planets travel around the Sun, but that their paths were ellipses.

Another teacher, Brahmagupta, was the first to say that zero was a number. This helped to give us the decimal system of numbers. Another great mathematician was named Bhaskara. In the 12th century, he worked on algebra and trigonometry. His work was taken to Arabia and later to Europe.

Decimals and place value

- Round numbers to a given number of significant figures.
- Use knowledge of place value to multiply and divide integers and decimals by 0.1 and 0.01.
- Estimate and multiply decimals by integers and decimals.
- Estimate and divide decimals by numbers with one decimal place.

Significant figures

The use of **significant figures** (s.f.) is a way of rounding a number. You have already rounded numbers using the number of decimal places.

The term 'significant' means important. So, a significant figure is an important figure.

> Here only the first significant figure is written, the rest are written as zeros.

> Only write the two most important figures. Write the rest as zeros.

> **LET'S TALK**
> What if the number had been halfway i.e. 6350? Would it be rounded down to 6300 or up to 6400?

Worked example

6387 people went to see a basketball game. Write 6387 to:

a 1 significant figure

The most important figure is the 6 as it is worth 6000.

Therefore 6387 written to 1 significant figure is 6000.

Check: To 1 significant figure 6387 is closer to 6000 than to 7000.

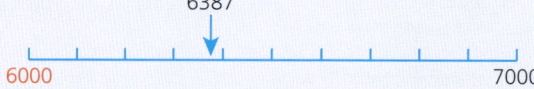

If the number had been, for example, 6732, then to 1 significant figure it would have been written as 7000, as it is closer to 7000 than 6000.

b 2 significant figures

The second most important figure is the 3 as it is worth 300.

Look at the number line.

The numbers 6300 and 6400 are the two numbers written to 2 significant figures either side of 6387.

6387 is closer to 6400, so 6387 written to 2 significant figures is 6400.

11 Decimals and place value

To round to a certain number of significant figures, look at the next digit to the one in question. If that digit is 5 or more, round up. If it is 4 or less, round down.

> **Worked example**
>
> A grain of sand is 0.0804 mm in diameter. Write this length to:
>
> a 1 significant figure
>
> Although the first two numbers are zero, they have no significance, therefore the 8 is the most significant number here. It is worth 8 hundredths.
>
> Checking on a number line whether to 1 significant figure 0.0804 is closer to 0.08 or 0.09.
>
> To 1 significant figure 0.0804 is therefore 0.08
>
>
>
> b 2 significant figures
>
> The first two zeros are ignored as they have no value. Therefore, the two most significant figures are the 8 and the 0 after it.
>
> We now know that 0.0804 to 2 significant figures must lie between 0.080 and 0.081.
>
> To work out which number it is closer to, we can draw a number line or simply look at the 3rd most significant number. If it is 5 or more then we round up.
>
> The 3rd most significant number is 4, therefore 0.0804 to 2 significant figures is 0.080.
>
> c The masses of two birds are recorded.
>
> The first bird has a mass of 1.32 kg correct to 3 s.f.
>
> The second bird has a mass of 2 kg correct to 1 s.f.
>
> Give the sum of their masses to an appropriate degree of accuracy.
>
> The masses are given to different degrees of accuracy. The answer should only be given to the degree of accuracy equal to that of the least accurate number, in this case 1 s.f.
>
> 1.32 + 2 = 2.32
>
> Therefore, the sum of their masses to an appropriate degree of accuracy is 2 kg.

In this case as the significant figure occurs after the decimal point, subsequent zeros can be left off. i.e. write 0.08 rather than 0.0800

Notice how the zero is included at the end indicating that the answer has been given to 2 significant figures. If it was written as 0.08, then this would only be given to 1 significant figure.

LET'S TALK
To 3 significant figures, what could be the lightest possible mass of the second bird? To 3 significant figures, what could its heaviest mass be?

SECTION 2

Exercise 11.1

1. Write each of the following numbers to 1 significant figure.
   a. 6320
   b. 785
   c. 93
   d. 4856
   e. 4.3
   f. 6.7
   g. 0.84
   h. 0.076

2. Write each of the following numbers to 2 significant figures.
   a. 6834
   b. 6874
   c. 4.62
   d. 7.38
   e. 52.84
   f. 46.39
   g. 0.8738
   h. 0.8783

3. A question in a book asks students to write 7058 to three significant figures.
   a. One student writes the answer 7050. What mistake has he made?
   b. Another student writes the answer 706. What mistake has she made?
   c. Write down the correct answer.

4. In the following statements:
   i) State which are correct and which are incorrect.
   ii) For those which are incorrect, explain why.
   a. In 2.315 the second most significant number is the 3.
   b. 2.315 rounded to 3 s.f. is 2.31.
   c. In the number 1832, the 8 is the most significant number as it is the biggest number.
   d. 1832 rounded to 2 s.f. is 18.
   e. In 0.0047 the 4 is the most significant number.
   f. 0.0047 rounded to 1 s.f. is 0.0050.
   g. 4.038 rounded to 2 s.f. is 4.04.
   h. 10.96 rounded to 1 s.f. is 10.
   i. 10.96 rounded to 2 s.f. is 11.
   j. 10.96 rounded to 3 s.f. is 10.9.

5. A plumber needs a piece of metal pipe which is 12.48 m in length. His supplier can only cut metal pipes to a length correct to 3 s.f.
   a. The plumber asks for a piece of metal pipe to be cut to a length of 12.5 m. Will this definitely be long enough? Justify your answer.
   b. What is the shortest length the plumber can ask for that guarantees he has got a long enough pipe?

6. A plane has a baggage mass restriction of 22 kg (to 2 s.f.) per person. One traveller checks in two bags. Their masses are 10 kg (to 1 s.f.) and 11 kg (to 2 s.f.).
   a. Calculate the combined mass of the two bags to an appropriate degree of accuracy.
   b. Is there a possibility that the traveller's baggage is heavier than allowed? Give a **convincing** argument for your answer.

7. A music venue has a seating capacity for exactly 3000 people.
   Tickets for the venue are sold by two different promoters. For one concert the promoters sell the following numbers of tickets:
   - Promoter 1   1600 (to 2 s.f.)
   - Promoter 2   1400 (to 2 s.f.)
   a. Is there a chance that too many tickets have been sold? Justify your answer.
   b. Based on these ticket sales, what is the maximum number of empty seats possible for this concert? Justify your answer.

Multiplying and dividing by numbers between 0 and 1

Multiplying and dividing by 0.1 and 0.01

You will already know from your work on decimals and fractions that decimals can be written as fractions (e.g. $0.5 = \frac{1}{2}$) and fractions can be written as decimals (e.g. $\frac{1}{5} = 0.2$). You will also be familiar with the fact that multiplying, say, by $\frac{1}{2}$ is the same as dividing by 2. Similarly dividing by $\frac{1}{2}$ is the same as multiplying by 2.

These relationships will help when it comes to multiplying or dividing by 0.1 and 0.01.

Worked example

a   Calculate the value of 15×0.1.

0.1 is equivalent to $\frac{1}{10}$, therefore the calculation can be written as $15 \times \frac{1}{10}$.

Multiplying by $\frac{1}{10}$ is equivalent to dividing by 10, therefore the calculation can be written as $15 \div 10$.

So $15 \times 0.1 = 15 \times \frac{1}{10}$
$= 15 \div 10$
$= 1.5$

b   Calculate the value of 24×0.01.

0.01 is equivalent to $\frac{1}{100}$, therefore the calculation can be written as $24 \times \frac{1}{100}$.

Multiplying by $\frac{1}{100}$ is equivalent to dividing by 100, therefore the calculation can be written as $24 \div 100$.

So $24 \times 0.01 = 24 \times \frac{1}{100}$
$= 24 \div 100$
$= 0.24$

c   Calculate the value of $15 \div 0.1$.

0.1 is equivalent to $\frac{1}{10}$, therefore the calculation can be written as $15 \div \frac{1}{10}$.

Dividing by $\frac{1}{10}$ is equivalent to multiplying by 10, therefore the calculation can be written as 15×10.

SECTION 2

So $15 \div 0.1 = 15 \div \frac{1}{10}$
$= 15 \times 10$
$= 150$

d Calculate the value of $24 \div 0.01$.

0.01 is equivalent to $\frac{1}{100}$, therefore the calculation can be written as $24 \div \frac{1}{100}$.

Dividing by $\frac{1}{100}$ is equivalent to multiplying by 100, therefore the calculation can be written as 24×100.

So $24 \div 0.01 = 24 \div \frac{1}{100}$
$= 24 \times 100$
$= 2400$

Exercise 11.2

1 By showing your method, work out the answer to each of the following multiplications:
- a 19×0.1
- b 48×0.1
- c 152×0.1
- d 8×0.1
- e 5012×0.1
- f 1×0.1
- g 170×0.01
- h 582×0.01
- i 49×0.01
- j 33×0.01
- k 5×0.01
- l 2×0.01

2 By showing your method, work out the answer to each of the following divisions:
- a $8 \div 0.1$
- b $14 \div 0.1$
- c $99 \div 0.1$
- d $1 \div 0.1$
- e $602 \div 0.1$
- f $4321 \div 0.1$
- g $2 \div 0.01$
- h $8 \div 0.01$
- i $29 \div 0.01$
- j $90 \div 0.01$
- k $482 \div 0.01$
- l $6521 \div 0.01$

3 A set of mathematical operations is shown on the cards below.

 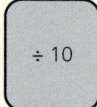 ÷ 100

Copy the calculations below. They show a number followed by a space and then an answer.
i) Choose one of the operations above to place in the space to make the calculation correct.
ii) Choose another of the operations above to place in the space to make the calculation correct.

- a 12   = 1.2
- b 102   = 10.2
- c 470   = 4.7
- d 36   = 360
- e 7   = 700
- f 123   = 12 300

11 Decimals and place value

> **Worked example**
>
> Estimate the answer to each of these multiplications.
>
> a 83×2.9
>
>   A good estimate would be $80 \times 3 = 240$
>
> b 83×0.29
>
>   A good estimate would be 80×0.3
>
>   This multiplication can be calculated in two ways:
>
>   i) As $80 \times 3 = 240$ then $80 \times 0.3 = 24$
>
>   0.3 is ten times smaller than 3, therefore the answer will also be ten times smaller.
>
>   ii) 80×0.3 can be split in to $80 \times 0.1 \times 3$
>
> $80 \times 0.1 = 8$
>
> $8 \times 3 = 24$
>
>   > Multiplying by 0.1 is the same as multiplying by $\frac{1}{10}$ and also the same as dividing by 10.
>
> c 83×0.029
>
>   A good estimate would be 80×0.03
>
>   This multiplication can be calculated in two ways:
>
>   i) As $80 \times 3 = 240$ then $80 \times 0.03 = 2.4$
>
>   0.03 is one hundred times smaller than 3, therefore the answer will also be one hundred times smaller.
>
>   ii) 80×0.03 can be split in to $80 \times 0.01 \times 3$
>
> $80 \times 0.01 = 0.8$
>
> $0.8 \times 3 = 2.4$
>
>   > Multiplying by 0.01 is the same as multiplying by $\frac{1}{100}$ and also the same as dividing by 100, i.e. $80 \times 0.01 = 80 \div 100 = 0.8$.

You can see that:
- Multiplying by a number greater than 1 increases the value.
- Multiplying by a number less than 1 decreases the value.

SECTION 2

Exercise 11.3

Estimate the answer to each of these multiplications.

In each question, use your estimate from part (a) to estimate the answers to parts (b) and (c).

1.   a 41×5 b 41×0.5 c 41×0.05
2.   a 59×6 b 59×0.6 c 59×0.06
3.   a 9.4×3 b 9.4×0.3 c 9.4×0.03
4.   a 7.93×4 b 7.93×0.4 c 7.93×0.04
5.   a 8.92×5 b 8.92×0.5 c 8.92×0.05
6.   a 12.39×6 b 12.39×0.6 c 12.39×0.06
7.   a 77.44×9 b 77.44×0.9 c 77.44×0.09
8.   a 42.12×8 b 42.12×0.8 c 42.12×0.08
9.   a 91.03×9 b 91.03×0.9 c 91.03×0.09
10.   a 0.93×2 b 0.93×0.2 c 0.93×0.02

Worked example

Estimate the answer to each of these divisions.

a $8.73 \div 3$

    A good estimate would be $9 \div 3 = 3$

b $8.73 \div 0.3$

    A good estimate would be $9 \div 0.3$.

    This division can be calculated in two ways:

    i)   As $9 \div 3 = 3$ then $9 \div 0.3 = 30$

       As 0.3 is ten times smaller than 3, the answer will be ten times bigger.

    ii) $9 \div 0.3$ can be split in to $9 \div 0.1 \div 3$

        $9 \div 0.1 = 90$

        $90 \div 3 = 30$

> Dividing by a smaller number will produce a bigger answer.

> Dividing by 0.1 is the same as dividing by $\frac{1}{10}$ and also the same as multiplying by 10.

11 Decimals and place value

Dividing by 0.01 is the same as dividing by $\frac{1}{100}$ and also the same as multiplying by 100, i.e.
$9 \div 0.01 = 9 \times 100$
$= 900$

c $8.73 \div 0.03$

A good estimate would be $9 \div 0.03$.

This division can be calculated in two ways:

i) As $9 \div 3 = 3$ then $9 \div 0.03 = 300$

As 0.03 is one hundred times smaller than 3, the answer will be one hundred times bigger.

ii) $9 \div 0.03$ can be split into $9 \div 0.01 \div 3$

$9 \div 0.01 = 900$

$900 \div 3 = 300$

LET'S TALK
Why is $9 \div 0.3$ split into $9 \div 0.1 \div 3$ and not $9 \div 0.1 \times 3$?

Exercise 11.4

Estimate the answer to each of these divisions.

In each question, use your estimate from part (a) to estimate the answers to parts (b) and (c).

1. a $37 \div 3$ b $37 \div 0.3$ c $37 \div 0.03$
2. a $7.7 \div 2$ b $7.7 \div 0.2$ c $7.7 \div 0.02$
3. a $29 \div 3$ b $29 \div 0.3$ c $29 \div 0.03$
4. a $8.1 \div 4$ b $8.1 \div 0.4$ c $8.1 \div 0.04$
5. a $37.65 \div 5$ b $37.65 \div 0.5$ c $37.65 \div 0.05$
6. a $17.72 \div 6$ b $17.72 \div 0.6$ c $17.72 \div 0.06$
7. a $42.19 \div 7$ b $42.19 \div 0.7$ c $42.19 \div 0.07$
8. a $87.56 \div 8$ b $87.56 \div 0.8$ c $87.56 \div 0.08$
9. a $18.36 \div 9$ b $18.36 \div 0.9$ c $18.36 \div 0.09$
10. a $50.86 \div 3$ b $50.86 \div 0.3$ c $50.86 \div 0.03$

You can see that:
- Dividing by a number greater than 1 decreases the value.
- Dividing by a number less than 1 increases the value; the closer the number is to 0, the greater the increase is.

SECTION 2

Exercise 11.5

For each of the questions 1 and 2, decide whether the statement is true or false. If false justify your answer.

1. It is given that $120 \times 3 = 360$, therefore:
   a $120 \times 0.3 = 36$
   b $120 \times 0.03 = 3.6$
   c $1200 \times 0.03 = 36$
   d $120 \div 0.3 = 360$
   e $120 \div 0.1 = 1200$
   f $12 \div 0.01 = 120$

2. It is given that $18.65 \div 0.05 = 373$.
   Write a related calculation that gives the following answers:
   a 37.3
   b 3730
   c 186.5
   d 50

3. A sheet of paper is 0.1 mm thick.
   a A ream of paper has 500 sheets. How thick would 1 ream of paper be?
   b A pile of paper has a thickness of 12 cm. How many sheets are in the pile?

A ream of paper is defined as 500 sheets of the same size and type of paper.

4. A paper manufacturer sells reams of paper according to mass.
   A ream of A4 paper weighs 2.2 kg (to 2 s.f.).
   The manufacturer charges $8.35 for a ream of this paper.
   The thickness of 1 sheet of this paper is 0.012 cm (to 2 s.f.).
   An order is shipped out which has a mass of 506 kg.
   a How much will the manufacturer charge for this paper?
   b If the paper was stacked up in one pile, how tall would it be?
      Show all of your working clearly.

▶ Now you have completed Unit 11, you may like to try the Unit 11 online knowledge test if you are using the Boost eBook.

12 Comparing and interpreting data

- Use knowledge of mode, median, mean and range to compare two distributions, considering the interrelationship between centrality and spread.
- Interpret data, identifying patterns, trends and relationships, within and between data sets, to answer statistical questions.

Averages and the range

Averages

> The word 'average' in everyday language is often used to represent the 'mean' but in mathematics, it can be used for the mean, median or mode.

You are already familiar with the most common types of statistical calculations that can be carried out on a set of data. These are the different types of average: the mean, the median and the mode.

- The **mean** is $\frac{\text{the sum of all the values}}{\text{the number of values}}$.
- The **median** is the middle value when the values have been arranged in order of size.
- The **mode** is the value which occurs the most often.
  With grouped data, the **modal group** is the group with the largest frequency.

The range

Another useful calculation that can be carried out on a set of data, and that you have encountered before, is to find the range. The range is the difference between the biggest value and the smallest value. It gives a measure of how spread out the data are.

Carrying out these statistical calculations on individual sets of data is fine; however they are particularly useful when comparing two sets of data.

SECTION 2

Worked example

A group of boys and a group of girls in the same class took the same test. The frequency tables show their marks out of 10.

Marks out of 10 (boys)	Frequency
1	1
2	0
3	0
4	2
5	2
6	3
7	4
8	2
9	0
10	1

Marks out of 10 (girls)	Frequency
1	0
2	0
3	0
4	2
5	4
6	6
7	1
8	1
9	0
10	0

a Calculate the mean, median, mode and range for the boys and girls.

	Boys	Girls
Mean	6.1	5.6
Median	6	6
Mode	7	6
Range	9	4

b The boys think they did better than the girls. Comment on this statement.

Although the boys' averages are slightly higher than the girls', they are only by a small amount. The range of the boys' results is much greater than that for the girls, which means the girls were much more consistent than the boys.

Therefore, the answer is not straightforward.

LET'S TALK
What values of mean, median and mode would be needed in order to say the boys definitely did better? Is there a precise number needed?

12 Comparing and interpreting data

Exercise 12.1

1 Two radio stations 'Listen FM' and 'Beatz' carry out a survey to find out the age of their listeners. The ages are summarised in the table opposite.

	Listen FM	Beatz
Mean	15.5	18
Median	14	17
Mode	17	15
Range	37	12

    a i) Which station has the listeners with the lowest mean age?
       ii) Which station has the listeners with the lowest modal age?
      iii) Both radio stations state that on average they have the younger listening audience. Explain why both can be correct.
    b Which radio station do you think has the younger audience? Justify your choice.
    c Which radio station has an audience with a bigger age range?
    d The youngest Beatz listener surveyed was 9 years old. What was the age of the oldest Beatz listener surveyed?

2 On her journey to work, Mala can choose either to drive on the motorway or to drive on smaller country lanes. She always leaves home at 8:00. She decides to time herself, in minutes, over a period of several weeks to see which route is the fastest. The results are in the table below:

	Mean	Median	Mode	Range	Fastest time (min)
Motorway	27	24	20	41	22
Country lanes	37	36	37	6	34

    a Mala says 'It is clear from the results that using the motorway is always quicker than using country lanes.' Explain whether her conclusion is correct.
    b Which route has the more consistent journey times?
    c One day, Mala has a meeting at work at 8:45. Assuming she still leaves home at 8:00, which route should she take if she wants to be certain to arrive on time?
    d i) One day, Mala has a meeting at work at 8:30. Assuming she still leaves home at 8:00, which route should she take?
       ii) Will she definitely arrive in time using the route you chose in part i?

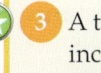

3 A tomato grower trials a new type of plant food to see if the number of tomatoes per plant increases as a result. Twenty tomato plants are given just water and another twenty are given the new plant food. He counts and records the number of tomatoes, of a certain size, that each plant produces. The two sets of data are given in the tables.

Plants given water only

Number of tomatoes per plant	30	31	32	33	34	35	36	37	38	39	40
Frequency	2	3	3	4	4	2	1	0	0	0	1

SECTION 2

Plants given new food

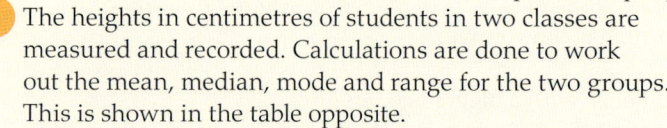

Number of tomatoes per plant	30	31	32	33	34	35	36	37	38	39	40
Frequency	3	3	1	1	0	0	0	4	4	1	3

a Calculate the mean, median and mode of the number of tomatoes produced per plant for each set of data.
b Calculate the range for each set of data.
c Comment on any differences that you found in parts (a) and (b) and decide whether the new food affects the number of tomatoes produced per plant.

4 The heights in centimetres of students in two classes are measured and recorded. Calculations are done to work out the mean, median, mode and range for the two groups. This is shown in the table opposite.
One of the classes is a Stage 7 class, the other a Stage 11 class. Decide which class is more likely to be the Stage 7 and which the Stage 11 class.
Justify your choice by using the data given in the table.

	Class A	Class B
Mean	161	147
Median	150	150
Mode	148	157
Range	35	15

5 The lengths in centimetres of leaves of two different plant types are measured and the results recorded. A summary of the data is presented in the table opposite.
For the statements below decide, justifying your answer, whether
　　i) it is true　　　　ii) it is not true
　　iii) it is not possible to tell whether it's true or not.
a On average, leaf X is bigger than leaf Y.
b There is a bigger spread of lengths in leaves of type Y.

	Type X	Type Y
Mean	14	14
Median	14	16
Mode	12	6
Range	5	18

Reliability of statistical calculations

Statistical calculations are carried out to summarise data. However, it is often said that statistics can be used to mislead. It is important to know as much as possible about the data and the methods of calculation before accepting any conclusions that are made.

The table highlights the strengths and weaknesses of the different types of statistical calculation you have met so far.

12 Comparing and interpreting data

Calculation	Strengths	Weaknesses
Mean	Takes all of the data into account Is the type of average understood by most people	Can be distorted by extreme results May not be a possible value for the data
Median	Takes all of the data into account Not affected by extreme values Easy to calculate	May not be a possible value for the data
Mode	Not affected by extreme values Easy to calculate Will always be a possible value for the data	The data may not have a mode or there may be several modes For grouped data, only a modal group can be given Does not take all of the data into account
Range	Easy to calculate Good for comparisons between data sets	Does not take all the data into account Can be distorted by extreme results

Worked example

The lifetimes (hours of continuous use in the same test) of ten batteries are shown below.

2   2   1   3   3   2   2   1   1   83

A manufacturer claims that 'on average' their batteries last 10 hours.

a   Is the claim true?

Calculating each of the three averages gives:

mean = 10 hours

median = 2 hours

mode = 2 hours

Strictly speaking, the claim is true. The manufacturer is taking the mean value as the average.

b   Is the claim misleading?

The claim is misleading because the mean calculation has been affected by the one extreme result. Neither the median nor the mode has been affected by this result.

c   What value for the average would you choose, and why?

Assuming that the intention is not to mislead, then either the median or the modal lifetime would be better.

LET'S TALK

'The average number of feet per human is less than two. Why are shoes always sold in pairs?'

Discuss the accuracy of this statement and also the answer to the question.

SECTION 2

Exercise 12.2

1 Two groups of ten students take the same mathematics test. One group is from a class which is set by ability, and the other is from a mixed ability class. The table shows their results.

Group A results	0	1	2	3	5	5	7	8	9	10
Group B results	5	5	5	5	5	5	5	5	5	5

  a Calculate the mean, median and modal results for each group.
  b Explain why none of the averages is useful for deciding which group is from which class.
  c Explain why calculating the range is helpful in this case.

2 A small company has seven employees and one manager. Their annual salaries are:

$6000	$8000	$7500	$10 000
$8000	$120 000	$7000	$8000

The company wishes to employ another person. The job advertisement states that the average salary in the company is approximately $21 800.
  a Comment on whether the claim in the advertisement is true.
  b Comment on whether you think the claim is misleading.

3 Two train companies publish data about how punctual their trains are. The data, which show how many minutes late trains arrive, are given in the table.

	Mean	Median	Mode	Range
Company A	10	3	1	54
Company B	8	8	8	4

  a Which train company is more reliable? Explain your answer.
  b Give a possible reason why the mean result for company A is so much higher than its median or mode.
  c Both companies want to advertise how punctual their trains are. Write a short sentence that each company might use for an advertising campaign.
  d Which company's trains do you think perform better? Justify your answer.

4 The table shows the midday temperatures (in °C) at two ski resorts over a 14-day period.

Midday temperature in resort X (°C)	−8	−6	−5	−5	−4	−3	−2	−1	0	1	1	4	4	4
Midday temperature in resort Y (°C)	−4	−3	−1	−1	−1	−1	−1	−1	0	0	0	0	1	1

  a Calculate the mean, median, mode and range for both resorts.
  b Resort X wishes to suggest that its temperatures are colder than those of resort Y. Write a sentence it could use to say this.
  c Resort Y wishes to suggest that its temperatures are colder than those of resort X. Write a sentence it could use to say this.

12 Comparing and interpreting data

Interpreting further data

Data are so central to 21st century society that it is important, not only to be able to carry out calculations with them and produce graphs from them, but also to be able to interpret them and understand the significance of what they show.

You have already looked at some comparisons between sets of data; this section will look at other types of interpretations.

> **Worked example**
>
> The graph shows the population of zebras and lions in a particular region over a long period of time.
>
>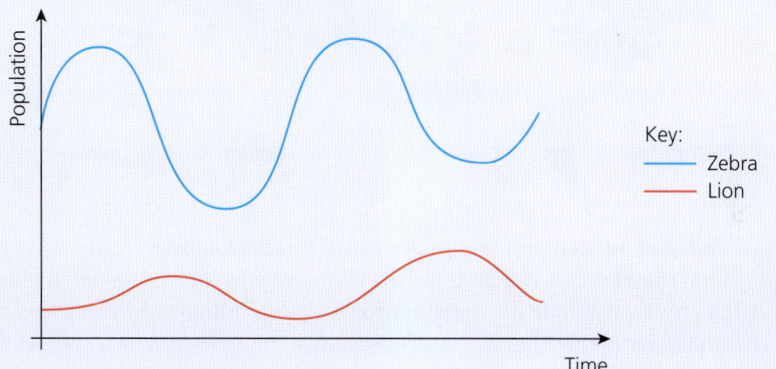
>
> **a** A farmer states that the graph shows that the populations of zebras and lions are dependent on each other. Explain whether you think this statement is true. Give a reasoned explanation for your answer by referring to the graph.
>
> The graph does appear to show a relationship between the two populations. As the population of zebras increases, the population of lions also tends to increase. As the lion population increases, this leads to a fall in the population of zebras; this leads, in turn, to a fall in the population of lions. There is a slight delay in the changes. The population of lions continues to increase once the population of zebras starts to fall; however, after a time the decrease in zebra numbers does affect the lion population.
>
> **b** Another region carries out a similar data collection. It finds that its graphs of zebra and lion populations are very different. What other factors could affect the shape of the graphs?
>
> Animal populations can be affected by many other factors, such as the existence of other types of predator, the impact of humans in the area through farming methods or urbanisation and the occurrence of diseases in the animal population.

SECTION 2

Exercise 12.3

1 The charts below show the percentage literacy rate for the different countries in the world and for two different age groups.

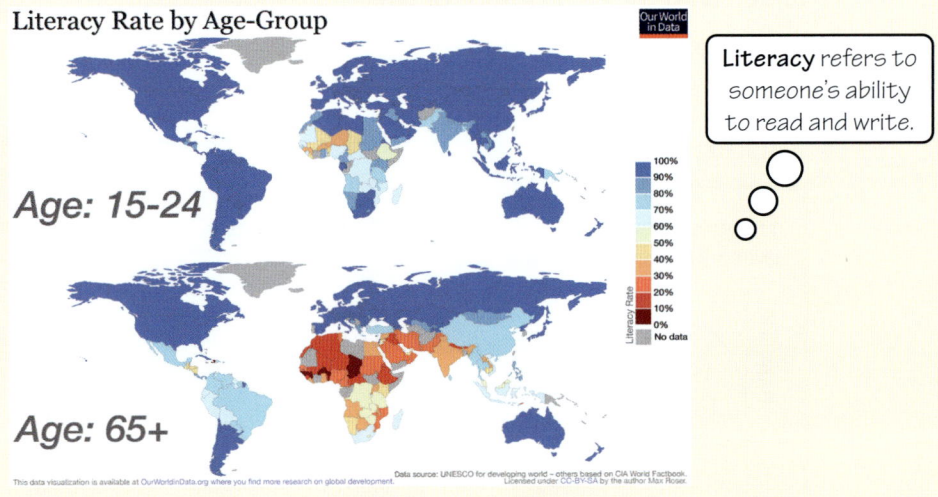

Literacy refers to someone's ability to read and write.

a What conclusions can you make about the data shown in the two charts?
b What reasons can you give for the difference in the results for the two age groups?
c If a chart was made for the age group 40–50, how would you expect it to differ from the two charts above? Justify your answer.

2 A hotel in a holiday resort in the Mediterranean produces a graph of the percentage of rooms occupied in different months of the year. This is shown in the graph:

To answer this question a knowledge of European holiday patterns and weather conditions in the Mediterranean will be helpful.

12 Comparing and interpreting data

a Explain possible reasons for the shape of this graph. Give at least three reasons to support your explanation.

b Another hotel nearby collects the same data and produces the following graph:

i)   State any similarities between the room occupancy levels in the two hotels.
ii) State any differences between the room occupancy levels in the two hotels.
iii) Give two factors which may have caused the differences between the two hotels.

c i) Draw a possible graph for the room occupancy of a hotel in a beach resort in the Southern hemisphere.
   ii) Justify the shape of your graph.

> **LET'S TALK**
> What place in the Southern hemisphere are you choosing? Discuss how that might affect the shape of your graph.

3 The graphs below show the age distribution of populations in two different regions of the world since 1950 to the present day.

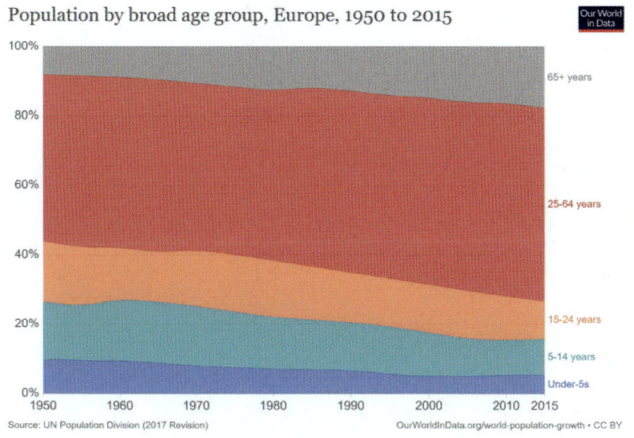

> **LET'S TALK**
> Spend time looking at the two graphs, discussing their shapes, similarities and differences and possible reasons for them before attempting the questions.

SECTION 2

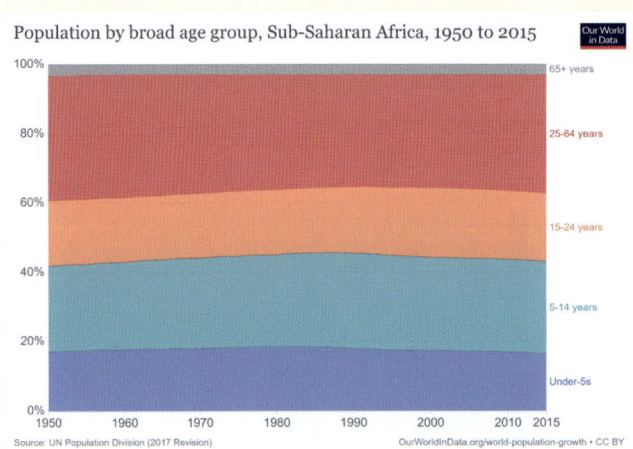

a i) Comment on any changes, over time, you have noticed on the graph for Europe.
   ii) Give a reason for why these changes may have happened.
b i) Comment on the significant differences between the graphs of the two regions.
   ii) Give a reason for why these differences may occur.
c You can access this data using the website 'Our World In Data' and searching for age structure.
   Use the website to produce a similar graph for your country and comment on any similarities and differences between your country and one of the regions shown above.

 Now you have completed Unit 12, you may like to try the Unit 12 online knowledge test if you are using the Boost eBook.

Transformation of 2D shapes

- Understand that the number of sides of a regular polygon is equal to the number of lines of symmetry and the order of rotation.
- Reflect 2D shapes and points in a given mirror line on or parallel to the x- or y-axis, or $y = \pm x$ on coordinate grids. Identify a reflection and its mirror line.
- Understand that the centre of rotation, direction of rotation and angle are needed to identify and perform rotations.
- Enlarge 2D shapes, from a centre of enlargement (outside or on the shape) with a positive integer scale factor. Identify an enlargement and scale factor.

Reflection and rotation

In Stage 7, you studied different types of transformations including reflection, rotation, enlargement and translation. This unit will look at the first three of those transformations in more detail.

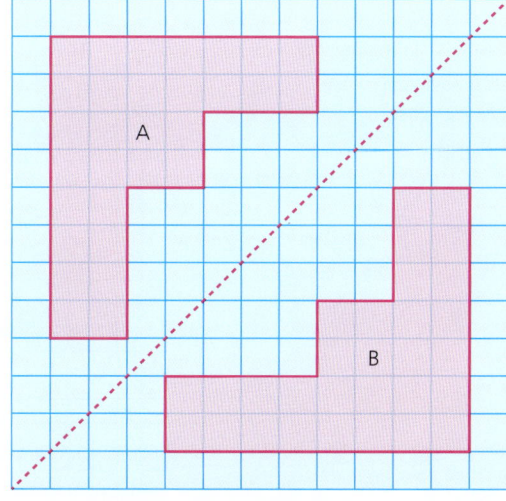

The **object** refers to the original position of the shape and its **image** is where it ends up after the transformation, in this case a reflection.

Object A is **reflected** in the **mirror line** to **map** on to **image** B. The mirror line becomes the line of symmetry of the final picture.

To describe a reflection, we must identify the position of the mirror line. If the diagram is drawn on a pair of axes, then its position can be described by stating the equation of the mirror line.

SECTION 2

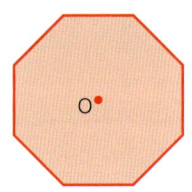

This regular octagon has **rotational symmetry** of **order** 8 when rotated about the **centre of rotation** O.

The order of rotational symmetry refers to how many times an object will look the same as it did at the start, when rotated a full 360°.

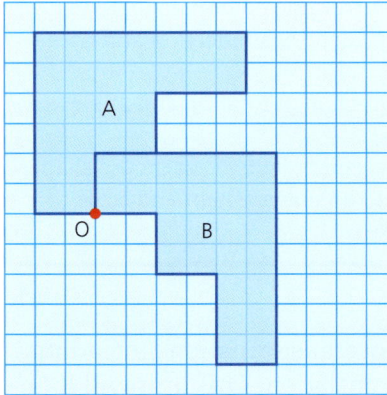

Object A is rotated 90° clockwise about the centre of rotation O, to map on to the image B.

To describe a rotation, we must identify the centre of rotation as well as the angle and the direction of rotation. If the diagram is drawn on a pair of axes, then the position of the centre of rotation can be stated using its coordinates.

Worked example

a On the axes opposite, shape P has been mapped on to shape Q as shown.

Describe the type of transformation fully.

Shape P has been reflected in the x-axis to map on to shape Q.

The equation of the mirror line is $y = 0$.

> **LET'S TALK**
> Why does the x-axis have the equation $y = 0$ and the y-axis have the equation $x = 0$?

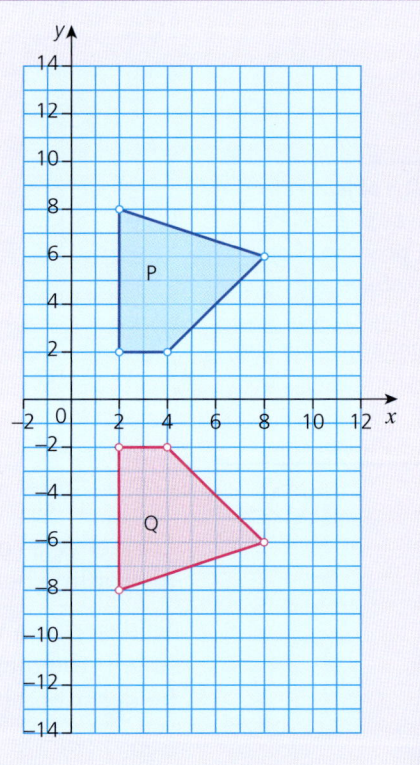

13 Transformation of 2D shapes

b On the axes shown, shape X is mapped on to shape Y.

Describe the transformation fully.

Shape X is rotated by 90° anti-clockwise to map on to shape Y.

The centre of rotation O has coordinates (2, 8).

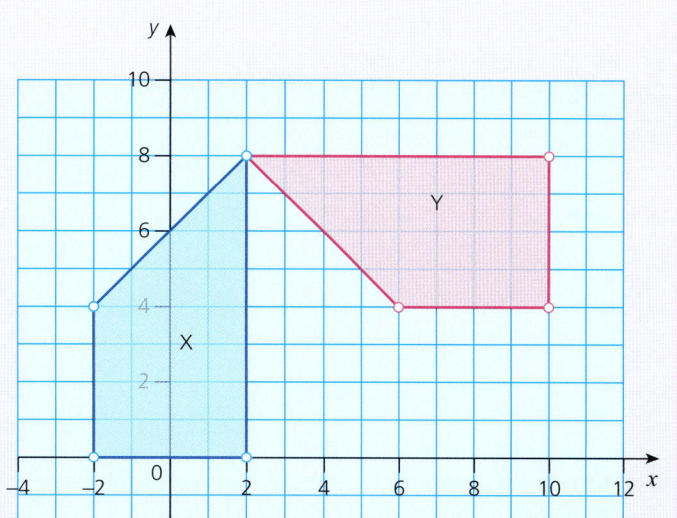

LET'S TALK

You will realise that 90° anti-clockwise is the same as 270° clockwise. It is usual therefore to give the angle and direction that is less than 180°.

LET'S TALK

If the rotation is 180° why does the direction not matter?

Exercise 13.1

1 The following are all regular polygons.

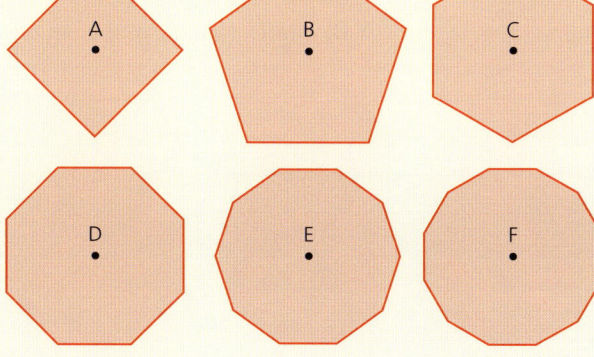

SECTION 2

a Copy and complete the table to **characterise** these regular polygons.

	Name	Number of sides	Number of lines of symmetry	Order of rotational symmetry
A				
B				
C				
D				
E				
F				

b Comment on your results.

2 A triangle A maps on to triangle B by a reflection as shown. Write down the equation of the mirror line.

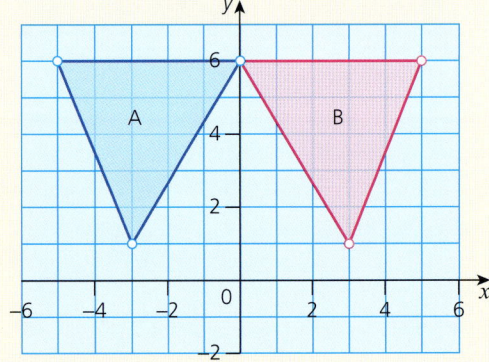

3 In the following questions shape X is mapped on to shape Y by a reflection. Give the equation of the mirror line in each case.

a

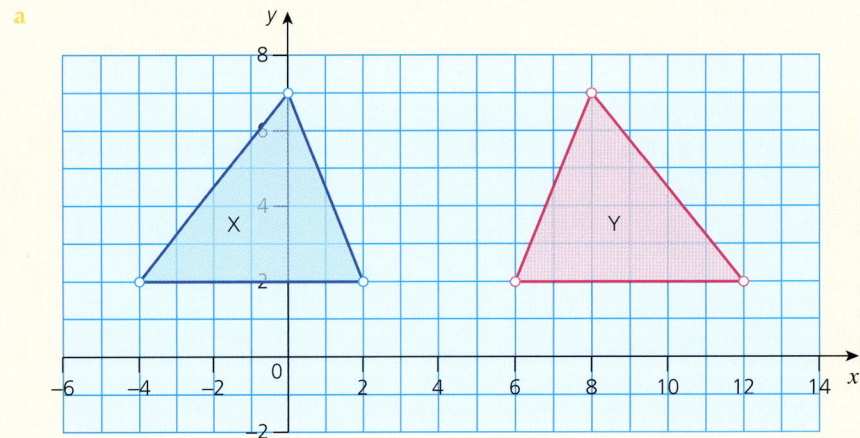

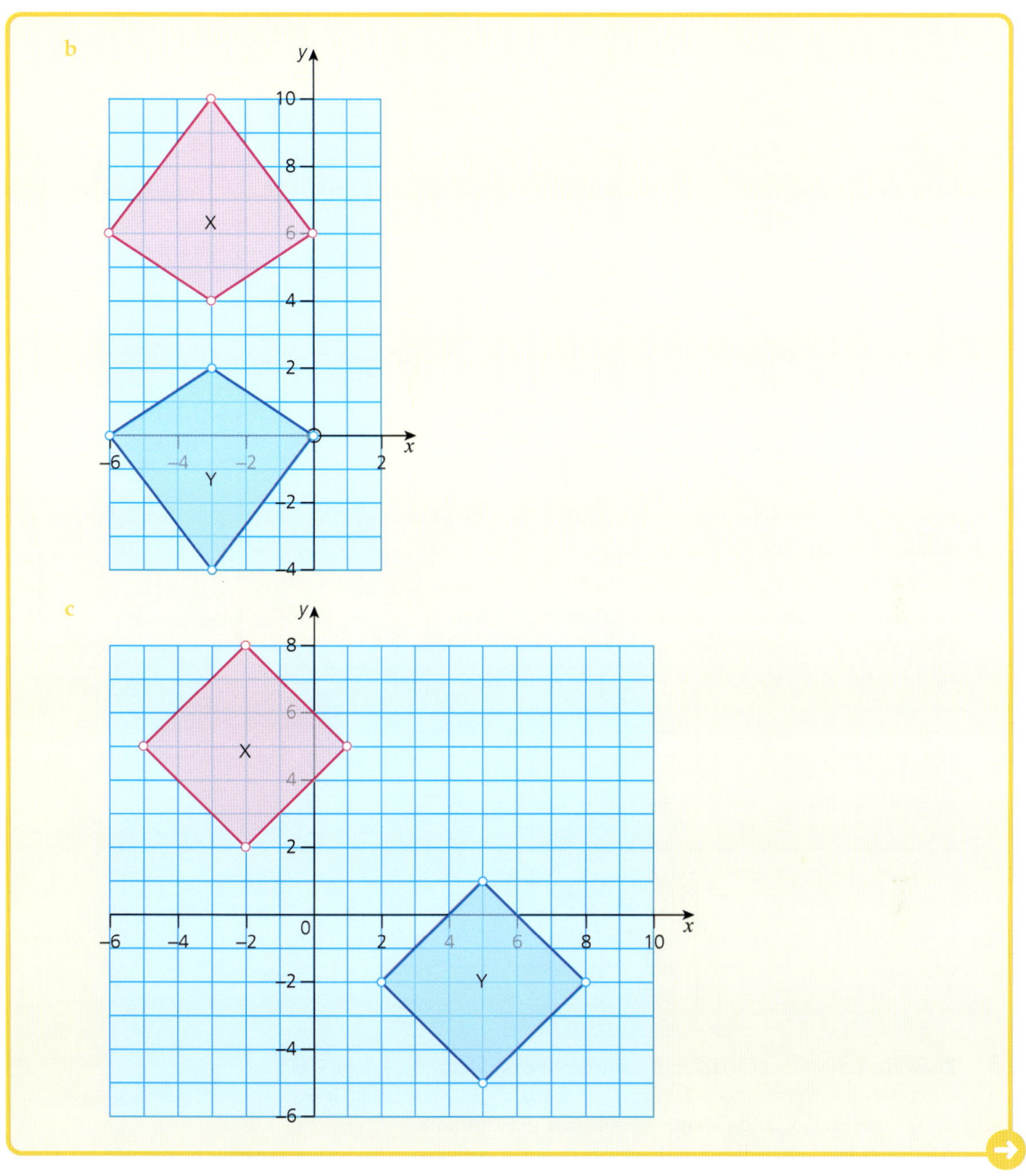

SECTION 2

d

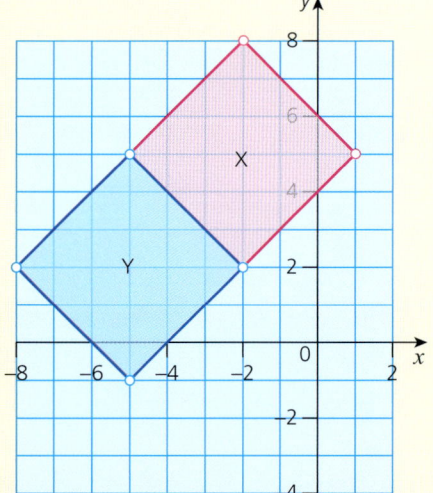

LET'S TALK
When an object is reflected in a vertical mirror line what happens to the x- and y-coordinates? What if the mirror line is horizontal?

Teachers are able to visit boost-learning.com where downloadable worksheets are available for the following question.

LET'S TALK
What does the coordinate (a, b) become if it is reflected in the line $y = x$? What about when it is reflected in the line $y = -x$?

4 In each of the following questions, use the worksheets provided by your teacher or copy the axes and the triangle LMN.
The equation of a mirror line is given each time.

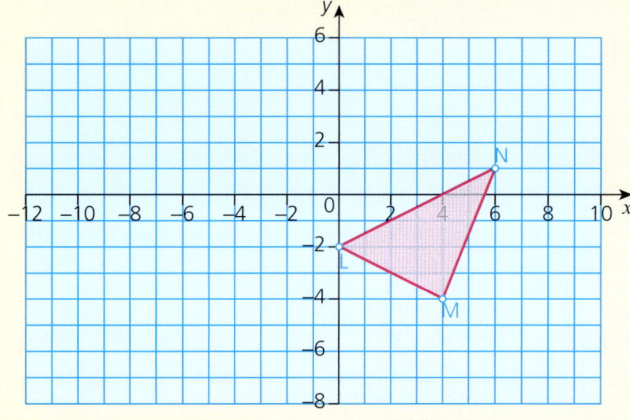

Draw the reflection of LMN in the mirror line and label the image L'M'N'.
 a $x = 0$ b $y = 1$ c $y = -2$ d $y = x$ e $y = -x$

5 The points A, B, C and D have the following coordinates:
A $(-1, -2)$ B $(-3, -1)$ C $(-1, 3)$ D$(1, 2)$
The coordinates are reflected in the mirror line $x = 1$ and map on to A', B', C' and D' respectively. The coordinates are given as follows:
A'$(3, -2)$ B'$(5, -2)$ C'$(3, 3)$ D'$(2, 2)$
 a Which of the mapped coordinates are incorrect? Justify your answer(s).
 b Give the correct mapped coordinates.

13 Transformation of 2D shapes

6 In each of the following questions the first point maps on to the second by a reflection. Deduce the equation of the mirror line each time.
  a   A(5, −2)   →   A'(5, 2)
  b   B(2, 6)   →   B'(−4, 6)
  c   C(−1, 3)   →   C'(−1, −1)
  d   D(0, 5)   →   D'(5, 0)

Rotation about a point

You already know that some shapes have rotational symmetry because they look exactly the same more than once when rotated 360° about a point.

An object can also be rotated about a different point to produce an image in a different place.

Worked example

a Draw the image Y when the object X is rotated by 90° anti-clockwise about the centre of rotation O.

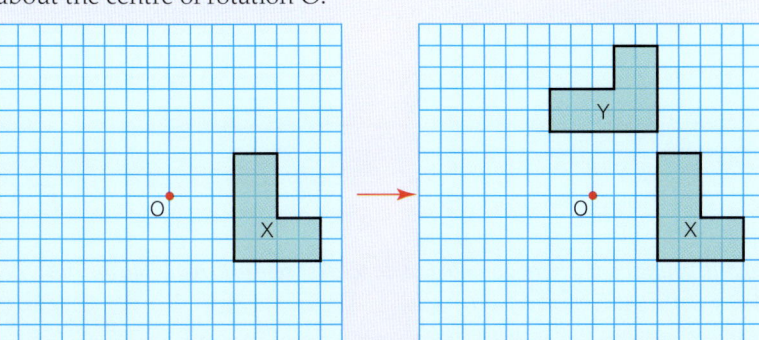

With a rotation, the object and image are **congruent**, i.e. exactly the same size and shape.

Every point on X is rotated by 90° anti-clockwise about O to the corresponding point on the image Y. The diagram below shows this for two of the points.

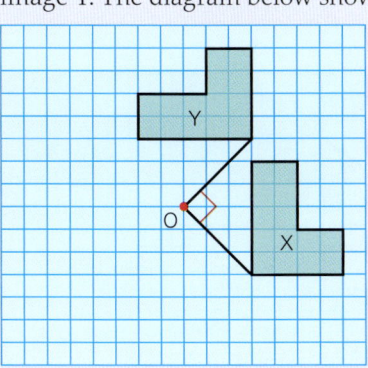
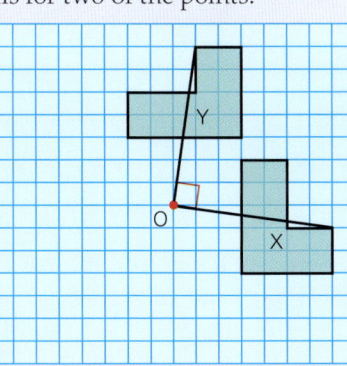

b Draw the image R when the object P is rotated by 180° about the centre of rotation O.

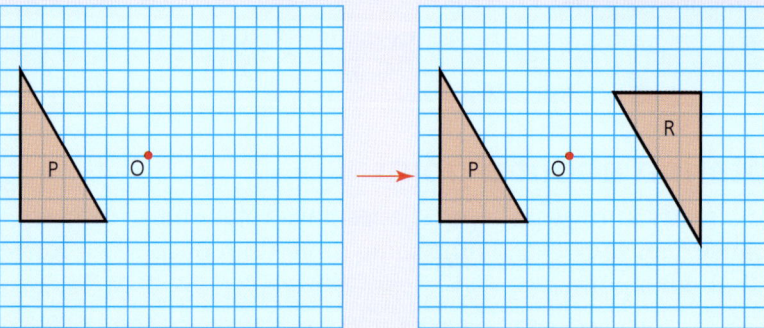

Once again, each point on the object is rotated by 180° about the centre of rotation, as shown below.

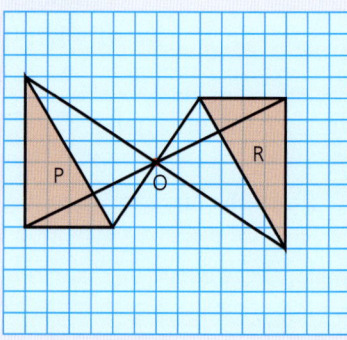

Exercise 13.2

Teachers are able to visit boost-learning.com where downloadable worksheets are available for the following questions.

In each of the following questions, use the worksheets provided by your teacher or copy the grid and the object A. Draw the image B when the object is rotated by the angle stated, about the centre of rotation O.

Use tracing paper to help you if necessary.

1
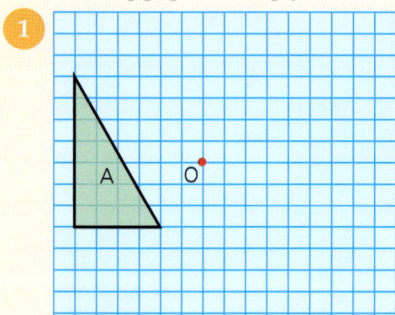
Rotation by 90° clockwise about O

2
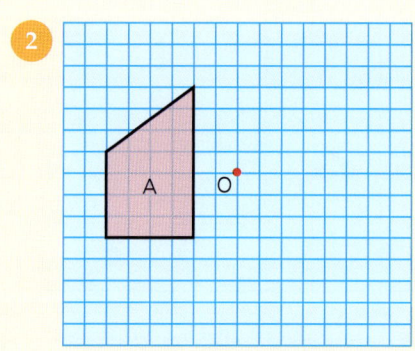
Rotation by 180° about O

13 Transformation of 2D shapes

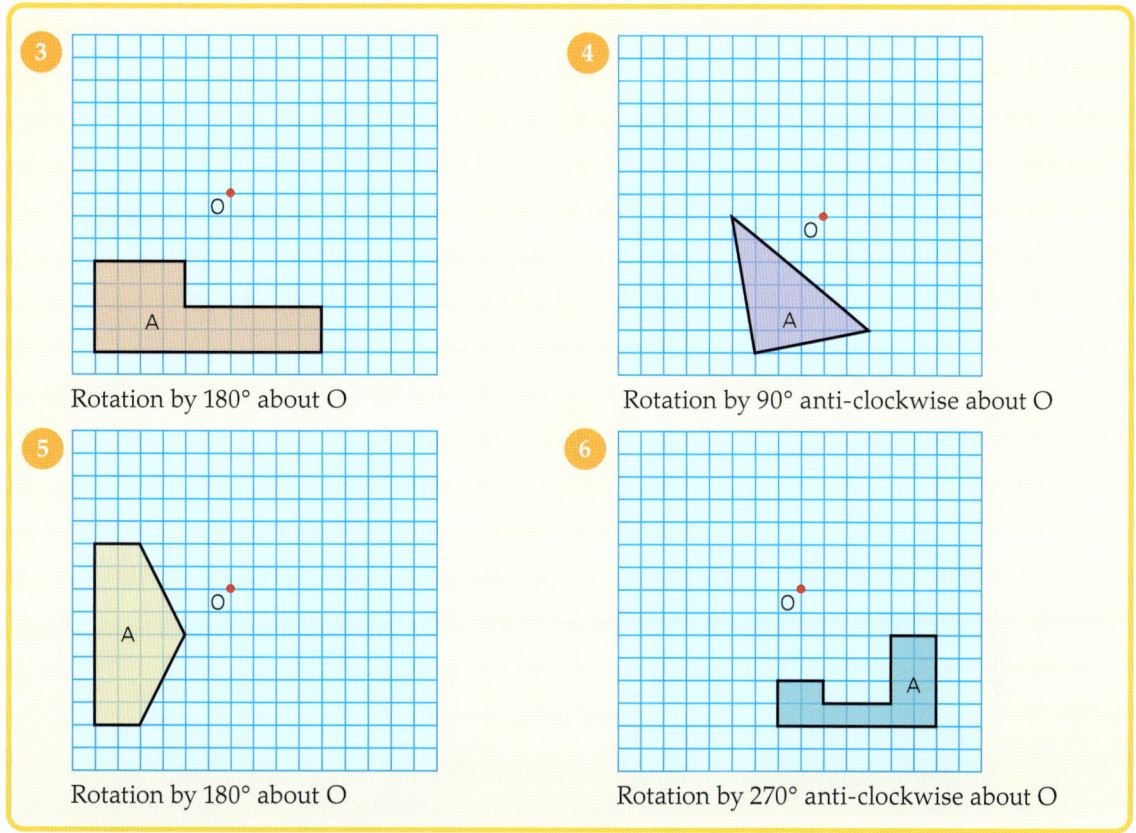

Additional transformations

All of the transformations you have met so far in this unit have been carried out as a single transformation. i.e. a reflection or a rotation. In Stage 7, you also encountered single translations. The following exercise will help to consolidate all the types of transformation you have encountered so far.

SECTION 2

> **Worked example**
>
> The object A undergoes two separate transformations:
> - a rotation by 90° clockwise about O
> - a reflection in the mirror line shown.
>
>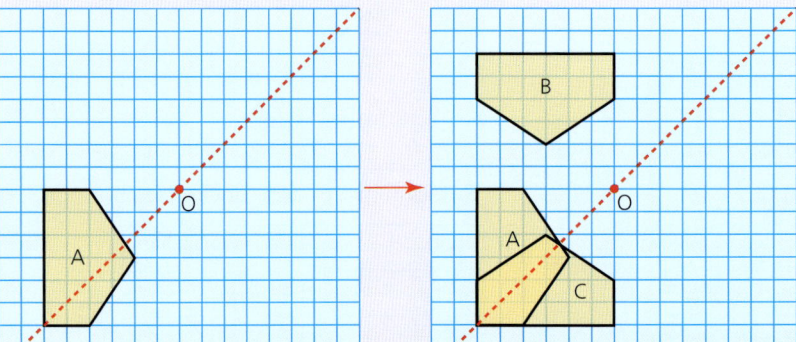
>
> Draw the images, labelling the image after the rotation B and the image after the reflection C.

Exercise 13.3

Teachers are able to visit boost-learning.com where downloadable worksheets are available for the following questions.

In each of the following questions, the object X undergoes two transformations. The first transformation maps X on to an image Y, the second maps X on to an image Z.

Use the worksheets provided by your teacher or copy each diagram on to squared paper and then draw each of the images, Y and Z, labelling them clearly.

1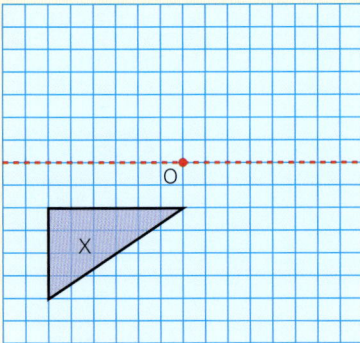

    a Reflection in the mirror line.
    b Rotation by 90° clockwise about O.

13 Transformation of 2D shapes

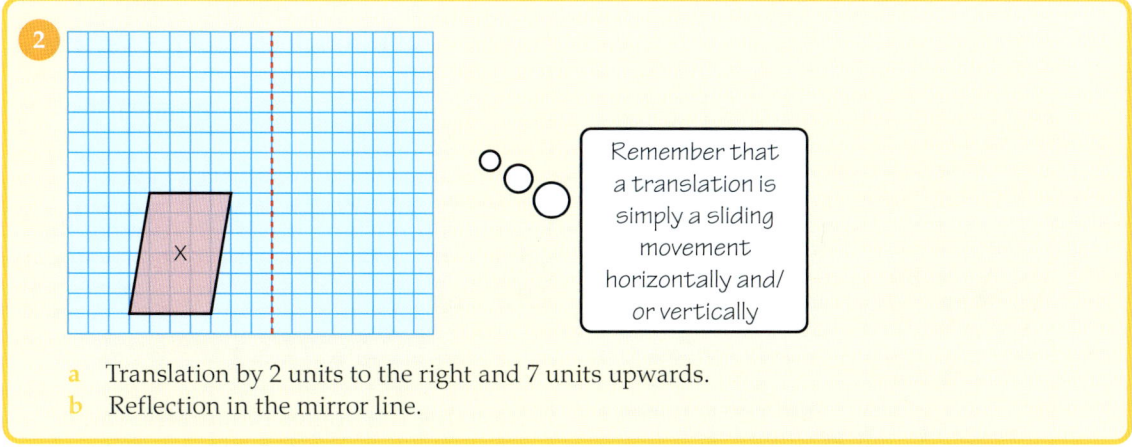

2

> Remember that a translation is simply a sliding movement horizontally and/or vertically

a Translation by 2 units to the right and 7 units upwards.
b Reflection in the mirror line.

Enlargement

With the transformations of reflection and rotation covered above the image produced is **congruent** to the original object. Both the angles and the lengths of the sides remain the same after the transformation.

However, with the transformation of **enlargement**, the image is mathematically **similar** to the object, but of a different size. The angles remain the same after the transformation, but the lengths of the sides change. The image can be larger or smaller than the original object. In this unit, we will only look at enlargements where the image is larger than the original object.

When describing an enlargement, two pieces of information need to be given. These are the position of the **centre of enlargement** and the **scale factor of enlargement**.

This diagram shows a triangle ABC enlarged to form triangle A'B'C'. The centre of enlargement is O and the scale factor of enlargement is 2.

SECTION 2

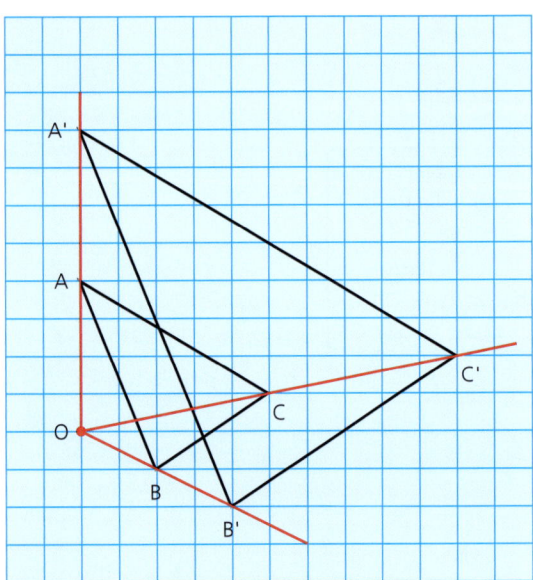

Lines can be drawn from the centre of enlargement through each of the vertices A, B and C. When extended, these lines also pass through the corresponding vertices A', B' and C'.

As the scale of enlargement is 2, each length in the triangle A'B'C' is double the length of the corresponding side in triangle ABC. That is,

$$\frac{A'B'}{AB} = \frac{A'C'}{AC} = \frac{B'C'}{BC} = 2$$

Notice that the angles remain the same.

The angle at A is the same size as the angle at A'.

Worked example

Enlarge the triangle XYZ by a scale factor of 2 and from the centre of enlargement O. Label the image X'Y'Z'.

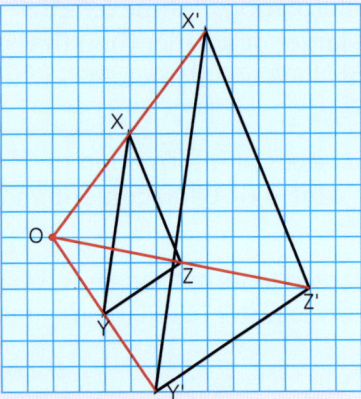

> When the enlargement is done on a grid, it is sometimes easier to count squares. For example, OX is 3 across and 4 up, so OX' is 6 across and 8 up.

To draw the enlargement, use the following steps.
- Draw lines from the centre of enlargement O through each of the vertices X, Y and Z.
- As the scale factor of enlargement is 2, draw the lines such that OX' is twice the length of OX, OY' is twice the length of OY and OZ' is twice the length of OZ.
- Join the ends of the lines OX', OY' and OZ' to form the enlarged shape X'Y'Z'.

13 Transformation of 2D shapes

Exercise 13.4

Teachers are able to visit boost-learning.com where downloadable worksheets are available for the following questions.

You may need squared paper for this exercise.

Use the worksheets provided by your teacher or copy each of the diagrams below onto squared paper. Enlarge each of the objects by the given scale factor and from the centre of enlargement O.

1 Scale factor of enlargement 2

2 Scale factor of enlargement 3

3 Scale factor of enlargement 2

4 Scale factor of enlargement 4

SECTION 2

In each of the following diagrams, the larger shape is an enlargement of the smaller one from the centre of enlargement O. Work out the scale factor of enlargement in each case.

5

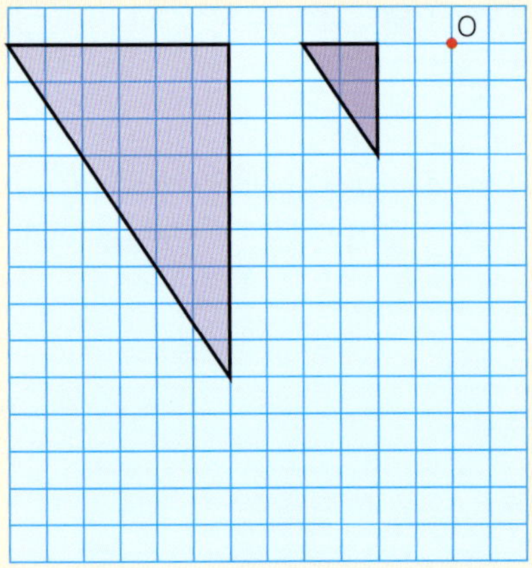

6

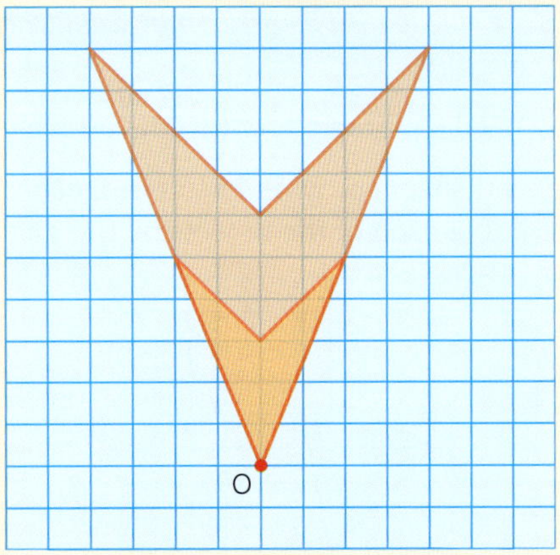

7

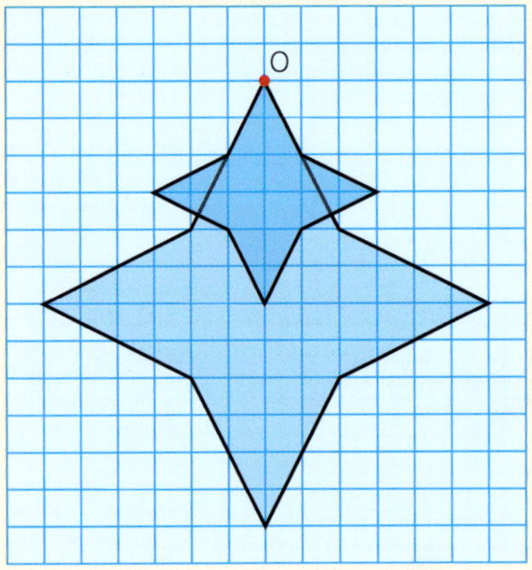

8

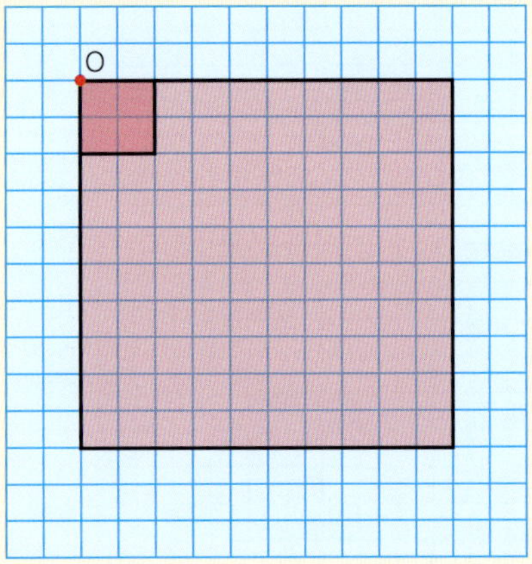

9 An image is projected on to a building.
The original image is 20 cm tall and 30 cm wide.
The actual building is 8 m tall and 22 m wide.
What is the maximum scale factor of enlargement of the image so that it appears completely on the building? Show all of your working clearly.

Now you have completed Unit 13, you may like to try the Unit 13 online knowledge test if you are using the Boost eBook.

14 Fractions and decimals

- Recognise fractions that are equivalent to recurring decimals.
- Estimate and subtract mixed numbers, and write the answer as a mixed number in its simplest form.
- Estimate and multiply an integer by a mixed number, and divide an integer by a proper fraction.
- Use knowledge of the laws of arithmetic and order of operations (including brackets) to simplify calculations containing decimals or fractions.
- Understand the relative size of quantities to compare and order decimals and fractions (positive and negative), using the symbols =, ≠, >, <, ≤ and ≥.

Terminating and recurring decimals

There are different types of decimal numbers.

You will already have encountered terminating decimals, which, as the name implies, have an end. For example, 0.32 or 4.8 or 3.6254.

A different type of decimal you will have used is called a recurring decimal: this has a number pattern that repeats after the decimal point. For example, 0.33333 etc. or 5.425425425 etc.

To show that a decimal is recurring, rather than writing etc. as above, a small 'dot' is placed above the repeating number(s). For example:

0.33333 etc. is written as 0.$\dot{3}$, implying that all subsequent numbers are 3.

> Another way of writing the recurring decimal 5.4$\dot{2}$$\dot{5}$ is 5.4$\dot{2}\dot{5}$, i.e. with a dot only above the first and last of the repeated numbers.

5.425425425 etc. is written as 5.4$\dot{2}\dot{5}$, which implies that all subsequent numbers repeat in the order 425.

Any decimal that is terminating or recurring is called a rational number because it can also be written as a fraction.

You have already covered in Stage 7 how to convert fractions into terminating decimals and vice versa. This unit will look at these again and also at identifying which fractions produce recurring decimals.

SECTION 2

Worked example

a Convert $\frac{3}{8}$ to a decimal. State whether it is a terminating or recurring decimal.

If the decimal equivalent of the fraction is not immediately obvious, then a division will be necessary:

$$\begin{array}{r} 0.375 \\ 8\overline{)3.^30^60^40 0} \end{array}$$

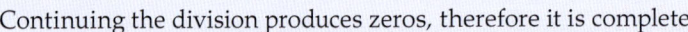

> If you already knew that $\frac{1}{8} = 0.125$, then you would probably have realised that $\frac{3}{8} = 0.375$ without the need for the division.

Continuing the division produces zeros, therefore it is complete.

$\frac{3}{8} = 0.375$, which is a terminating decimal.

b Convert $\frac{3}{11}$ to a decimal and state whether it is a terminating or recurring decimal.

To convert $\frac{3}{11}$ to a decimal divide 3 by 11.

$$\begin{array}{r} 0.2727 \\ 11\overline{)3.^30^80^30^8 0} \end{array}$$

As the remainder each time is alternating between 3 and 8, the answers will also repeat themselves, in this case between 2 and 7.

Therefore $\frac{3}{11} = 0.2\dot{7}$ which is a recurring decimal.

LET'S TALK
From this result can you deduce, without the need for a division, the decimal equivalent of $\frac{1}{11}$ and $\frac{5}{11}$?

Exercise 14.1

1 i) For each of the following fractions, state or calculate their decimal equivalent.
   ii) State whether the fraction is terminating or recurring.

   a $\frac{1}{3}$ d $\frac{5}{12}$
   b $\frac{2}{5}$ e $\frac{4}{9}$
   c $\frac{3}{4}$ f $\frac{2}{15}$

2 i) For each of the following fractions, state or calculate their decimal equivalent.
   ii) State whether the fraction is terminating or recurring.

   a $\frac{2}{9}$ d $\frac{1}{7}$
   b $\frac{2}{125}$ e $\frac{3}{7}$
   c $\frac{3}{20}$ f $\frac{3}{14}$

3 a Use the table to **classify** the fractions from questions 1 and 2.

Fractions that are equivalent to terminating decimals	Fractions that are equivalent to recurring decimals

   b Identify one property of all the fractions which produce terminating decimals.

130

14 Fractions and decimals

    c Identify one property of all the fractions which produce recurring decimals.
    d Write two more fractions in each side of the table. Explain how you decided to which side of the table they each belonged.

4 a Using division, calculate the decimal equivalent of $\frac{2}{3}$.
    b A calculator will give the answer to $2 \div 3$ as 0.6666666667.
       Comment on the reason why the calculator's answer differs from yours.

5 i) Decide whether each of the following **conjectures** is always true, sometimes true or never true.
    ii) For cases which are sometimes true, give a **convincing** explanation for your answer by giving an example of when it is and when it is not true.
      a A fraction where the denominator is a prime number will produce a recurring decimal.
      b A fraction where the denominator is a prime number other than 2 or 5 will produce a recurring decimal.
      c A fraction where the denominator is not a prime number will produce a terminating decimal.
      d A fraction where the denominator is a multiple of a prime number will produce a recurring decimal.
      e A fraction with a denominator of 5 will produce a recurring decimal.
      f A fraction with a denominator of 3 will produce a recurring decimal.

6 The fraction $\frac{22}{7}$ is often used as an approximation for π.
    a Using the internet as a reference write down π to 15 decimal places.
    b i) Using division, work out $\frac{22}{7}$.
       ii) Comment on any similarities and differences between the two values. Why might it be an advantage to use $\frac{22}{7}$ as an approximation for π?

> **LET'S TALK**
> By looking at π to 100 decimal places, can you see whether it is a recurring or a terminating decimal or neither?

Multiplications involving decimals can often look quite difficult. However, there are two key rules of multiplication which can make some of the calculations easier to do.

The order in which a multiplication is carried out will not affect the answer. For example:

$$4 \times 2 \times 7 = 2 \times 7 \times 4 = 7 \times 4 \times 2$$

Therefore changing the order can sometimes make the calculations easier.

Another rule of multiplication involves the use of brackets. For example:

$$4 \times (3 + 8) = 4 \times 3 + 4 \times 8$$

In other words a number multiplying a sum or subtraction inside a pair or brackets will give the same answer as multiplying each of the numbers inside the bracket by the number outside first. This too can sometimes make the calculation easier.

SECTION 2

> **Worked example**
>
> a Without a calculator work out $4 \times 8.2 \times 2.5$.
>   This can be rearranged as $4 \times 2.5 \times 8.2 = 10 \times 8.2 = 82$
> b Work out 5.2×9.9.
>   Treating 9.9 as $10 - 0.1$, makes the calculation easier.
> $5.2 \times (10 - 0.1) = (5.2 \times 10) - (5.2 \times 0.1)$
> $ = 52 - 0.52$
> $ = 51.48$
>
>  As 4×2.5 produces a nice answer, doing this first simplifies the calculation.

Exercise 14.2

1. A student wants to multiply the following numbers together: $0.37 \times 0.2 \times 5$. He starts by multiplying 0.37×0.2.
   a Explain how this method could be **improved**.
   b Carry out the multiplication in a more efficient way and justify your choice.
2. Work out the answer to the following:
   a $5.5 \times 7 \times 2$
   b $2.5 \times 9.3 \times 4$
   c $5 \times 0.62 \times 20$
   d $0.71 \times 0.4 \times 5$
3. Calculate the answer to the following:
   a 4.4×10.1
   b 35×9.8
   c 0.4×2.2
   d 11×6.2
   e 0.8×5.5

Mixed numbers and improper fractions

There are three types of fraction. These are:
- **Proper fractions**, where the numerator is smaller than the denominator, for example $\frac{3}{7}$.
- **Improper fractions**, where the numerator is bigger than the denominator, for example $\frac{8}{5}$.
- **Mixed numbers**, where there is a mixture of a whole number and a proper fraction, for example $1\frac{2}{3}$.

By their very definition, an improper fraction, because the **numerator** is bigger than the **denominator**, will always have a value greater than 1.

Mixed numbers, because they are a combination of a whole number and a proper fraction, will therefore also have a value greater than 1.

Therefore, it follows that an improper fraction can be written as a mixed number and vice versa. For example, $\frac{9}{4} = 2\frac{1}{4}$ and $3\frac{1}{3} = \frac{10}{3}$.

14 Fractions and decimals

> **Worked example**
>
> a Convert $\frac{12}{5}$ to a mixed number.
>
> This can be read as twelve fifths. Visually, this can be represented as follows:
>
>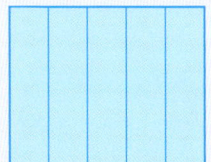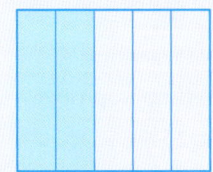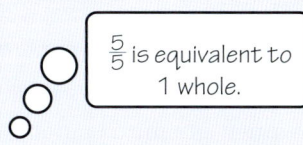
>
> $\frac{5}{5}$ is equivalent to 1 whole.
>
> Each whole square has been divided into fifths, showing that is equivalent to one whole.
>
> By shading in twelve fifths it can be seen that $\frac{12}{5} = 2\frac{2}{5}$.
>
> An alternative method is to divide 12 by 5, see how many times 5 goes in to 12 and what the remainder is. i.e. $12 \div 5 = 2$ and remainder **2**.
>
> Therefore $\frac{12}{5} = 2\frac{2}{5}$ ← The remainder
>
> Whole number of times 5 goes in to 12
>
> b Convert $4\frac{1}{3}$ to an improper fraction.
>
> This can be read as four wholes and a third of a whole, which visually can be represented as follows:
>
>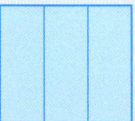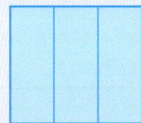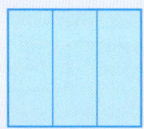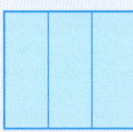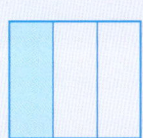
>
> As each whole has been split into thirds, $4\frac{1}{3}$ is equivalent to thirteen thirds $4\frac{1}{3} = \frac{13}{3}$.
>
> Another method is to multiply the whole number by the denominator and add the numerator.
>
> $4\frac{1}{3} = \frac{13}{3}$

Calculations involving mixed numbers

There are a number of different ways to carry out additions and subtractions involving mixed numbers. Which method you choose to use is largely up to you; however, being aware of different methods will enable you to choose a method which makes the calculation easier.

SECTION 2

Worked example

a Work out the following subtraction.

Method 1:

$3\frac{2}{5} - 1\frac{1}{5}$

This can rewritten as improper fractions:

i.e. $3\frac{2}{5} - 1\frac{1}{5} = \frac{17}{5} - \frac{6}{5} = \frac{11}{5}$, as $\frac{11}{5}$ is an improper fraction it can be written as a mixed number as $2\frac{1}{5}$

Therefore, $3\frac{2}{5} - 1\frac{1}{5} = 2\frac{1}{5}$

Method 2:

$3\frac{2}{5} - 1\frac{1}{5}$ can be written as $\left(3 + \frac{2}{5}\right) - \left(1 + \frac{1}{5}\right)$

Removing the brackets gives: $3 + \frac{2}{5} - 1 - \frac{1}{5}$

This can be rearranged to make an easier calculation as $3 - 1 + \frac{2}{5} - \frac{1}{5} = 2\frac{1}{5}$ as before.

> Note that the denominators here are the same, so the calculation is a bit easier.

> Note that the negative sign in front of the second bracket is subtracting everything inside it.

b Work out the following subtraction.

$5\frac{1}{6} - 2\frac{5}{6}$

Method 1:

Converting to improper fractions gives:

$5\frac{1}{6} - 2\frac{5}{6} = \frac{31}{6} - \frac{17}{6} = \frac{14}{6}$

$\frac{14}{6}$ written as a mixed number becomes $2\frac{2}{6}$

Therefore, $5\frac{1}{6} - 2\frac{5}{6} = 2\frac{1}{3}$

> Note that here, too, the denominators of both fractions are the same.

> Wherever possible fractions should be simplified. Here $\frac{2}{6}$ can be simplified to $\frac{1}{3}$.

Method 2:

$5\frac{1}{6} - 2\frac{5}{6}$ can, as in example (a), be written using brackets: 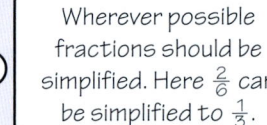 $\left(5 + \frac{1}{6}\right) - \left(2 + \frac{5}{6}\right)$

However, when removing the brackets there is a complication as shown:

$5 + \frac{1}{6} - 2 - \frac{5}{6}$ can be rewritten as $5 - 2 + \frac{1}{6} - \frac{5}{6}$

However, subtracting $\frac{5}{6}$ from $\frac{1}{6}$ will produce a negative value which complicates the calculation.

Therefore, to avoid this problem the first mixed number can be written differently:

$5\frac{1}{6} = 4 + 1\frac{1}{6} = 4\frac{7}{6}$

The original calculation using brackets is now, $\left(4 + \frac{7}{6}\right) - \left(2 + \frac{5}{6}\right)$

Removing the brackets gives, $4 + \frac{7}{6} - 2 - \frac{5}{6}$ and this can be rewritten as:

$4 - 2 + \frac{7}{6} - \frac{5}{6} = 2\frac{2}{6} = 2\frac{1}{3}$ as before.

c   Work out the following subtraction.

$4\frac{1}{3} - 3\frac{3}{4}$

Method 1:

Converting to improper fractions gives:

$\frac{13}{3} - \frac{15}{4}$

> Note that here the denominators of the fractions are different.

As the fractions have different denominators, we need to write them as equivalent fractions with the same denominator. This involves finding the lowest common multiple of 3 and 4, i.e. 12.

Therefore, $\frac{13}{3} - \frac{15}{4} = \frac{52}{12} - \frac{45}{12} = \frac{7}{12}$

Therefore, $4\frac{1}{3} - 3\frac{3}{4} = \frac{7}{12}$

Method 2:

$4\frac{1}{3} - 3\frac{3}{4}$ written using brackets gives $\left(4 + \frac{1}{3}\right) - \left(3 + \frac{3}{4}\right)$

As the fraction part of the second bracket is bigger than that in the first bracket, we will have the same problem as in example (b) in that we will get a negative fraction.

Therefore, the first bracket can be rewritten as:

$\left(3 + 1\frac{1}{3}\right) - \left(3 + \frac{3}{4}\right) = \left(3 + \frac{4}{3}\right) - \left(3 + \frac{3}{4}\right)$

Removing the brackets gives:

$3 + \frac{4}{3} - 3 - \frac{3}{4}$ which simplifies to $\frac{4}{3} - \frac{3}{4}$

Rewriting as equivalent fractions with the same denominator gives $\frac{16}{12} - \frac{9}{12} = \frac{7}{12}$.

Therefore, $4\frac{1}{3} - 3\frac{3}{4} = \frac{7}{12}$ as before.

Exercise 14.3

1. Look at both methods in the three worked examples.
   Which method did you find easiest in each of the examples?
   Justify your choices.

2. Work out the following. Give each answer in its simplest form.

   a $4\frac{1}{8} + 2\frac{3}{8}$

   b $3\frac{4}{5} + 5\frac{2}{5}$

   c $2\frac{3}{7} + 1\frac{4}{5}$

   d $5\frac{5}{12} - 3\frac{1}{12}$

   e $6\frac{1}{8} - 3\frac{5}{8}$

   f $4\frac{3}{8} - 3\frac{5}{9}$

   > Note that the abbreviation for hectares is ha.

3. A farmer has a plot of land with an area of 10 hectares.
   He uses $3\frac{1}{5}$ ha for root vegetables, $2\frac{1}{4}$ ha for corn and the remainder for strawberries.
   What area of the land is used for strawberries?

4. The diagram shows some weighing scales with masses on each side.

   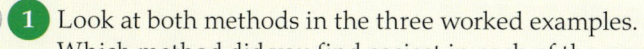

SECTION 2

a Do the scales balance? Justify your answer by estimation.
b How much mass must be removed from the heaviest side to balance the scales? Show all of your working clearly.

5 Five masses are given below:

$1\frac{2}{3}$ kg $2\frac{3}{4}$ kg $5\frac{1}{6}$ kg $2\frac{7}{12}$ kg $3\frac{1}{3}$ kg

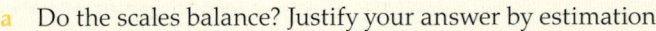

LET'S TALK
How does finding the total of all five masses help solve this problem?

Which masses should be placed on each side of the scales so that they balance? Show clearly how you reached your answer.

Multiplication and division involving mixed numbers

In Stage 7, you learned how to multiply and divide proper fractions.

Working with improper fractions and mixed numbers is no different as the rules of multiplication and division do not change.

Worked example

a Work out the following multiplication.

$5 \times 2\frac{3}{4}$

There are several ways in which this can be worked out.

Method 1:

5 can be written as a fraction as $\frac{5}{1}$

$2\frac{3}{4}$ can be written as an improper fraction as $\frac{11}{4}$

So, $5 \times 2\frac{3}{4} = \frac{5}{1} \times \frac{11}{4} = \frac{55}{4}$

$\frac{55}{4} = 13\frac{3}{4}$

Therefore, $5 \times 2\frac{3}{4} = 13\frac{3}{4}$

Method 2:

$5 \times 2\frac{3}{4}$ can also be written as $5(2 + \frac{3}{4})$

Expanding the bracket gives the following calculation:

$5 \times 2 + 5 \times \frac{3}{4} = 10 + \frac{15}{4}$

$= 10 + 3\frac{3}{4} = 13\frac{3}{4}$ as before.

KEY INFORMATION
A number multiplying terms inside a bracket, multiplies all the terms in the bracket.

b Work out the following division.

$4 \div \frac{1}{3}$

This implies finding how many 'thirds' go in to four and can be visualised as shown.

Each whole has been divided into thirds. As can be seen, there are 12 thirds in four wholes.

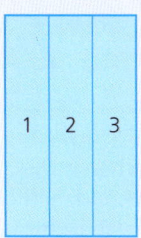 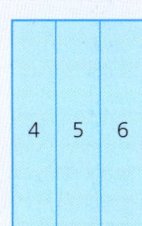 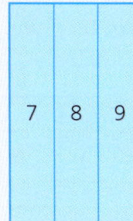 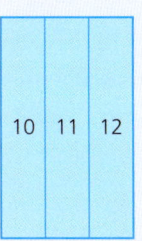

Therefore, $4 \div \frac{1}{3} = 12$.

But $4 \times 3 = 12$ therefore $4 \div \frac{1}{3} = 4 \times \frac{3}{1} = 12$.

Therefore, dividing by a fraction is the same as multiplying by its reciprocal.

The reciprocal of a number is 1 divided by that number.

For example, the reciprocal of 5 is $\frac{1}{5}$. Similarly, the reciprocal of $\frac{1}{3}$ is $\frac{1}{\left(\frac{1}{3}\right)} = 3$

c Work out the following calculation.

$5\frac{1}{4} - 1\frac{3}{4} \times 2\frac{1}{5}$

The multiplication part is done first.

Therefore, $1\frac{3}{4} \times 2\frac{1}{5} = \frac{7}{4} \times \frac{11}{5} = \frac{77}{20}$

The calculation is now:

$5\frac{1}{4} - \frac{77}{20} = \frac{21}{4} - \frac{77}{20}$ (convert the mixed number to an improper fraction)

$= \frac{105}{20} - \frac{77}{20}$ (write as equivalent fractions with the same denominator)

$= \frac{28}{20}$

$= \frac{7}{5} = 1\frac{2}{5}$ (write the improper fraction as a mixed number)

d Work out the following multiplication:

3.8×10.1

This can be estimated first as $4 \times 10 = 40$.

> **LET'S TALK**
> How can the division $x \div \frac{1}{y}$ be written as a multiplication?

> **KEY INFORMATION**
> When doing calculations with numerous fractions, the order of operations still follows the BIDMAS order.

14 Fractions and decimals

SECTION 2

This can be calculated in several ways using methods you have already covered.

Method 1: convert to fractions

$3.8 \times 10.1 = 3\frac{8}{10} \times 10\frac{1}{10}$

$= \frac{38}{10} \times \frac{101}{10} = \frac{3838}{100}$ (convert to improper fractions and multiply)

$= 38\frac{38}{100}$ (write as a mixed number)

$= 38.38$ (if the answer is to be given as a decimal)

> 38×101 is the same as
> $38 \times 100 + 38 \times 1 = 3800 + 38 = 3838$

Method 2: split the decimals to make the calculation easier

$3.8 \times (10 + 0.1) = 3.8 \times 10 + 3.8 \times 0.1$

$= 38 + 0.38$

$= 38.38$

> Multiplying by 0.1 is the same as dividing by 10.

> **LET'S TALK**
> When is this second method most useful?

e Insert one of the following symbols <, > or = when comparing both expressions below.

$\left(3\frac{1}{4} - 1\frac{1}{2}\right)^2 \ldots 2 \div \frac{3}{4}$

The value of each side needs to be calculated in order to compare them.

Calculating the left-hand side:

Using BIDMAS the bracket needs to be worked out first before being squared:

$3\frac{1}{4} - 1\frac{1}{2} = \frac{13}{4} - \frac{3}{2}$ (converting to improper fractions)

$= \frac{13}{4} - \frac{6}{4}$ (writing as equivalent fractions with the same denominator)

$= \frac{7}{4}$

Squaring the result is done next:

$\left(\frac{7}{4}\right)^2 = \frac{49}{16} = 3\frac{1}{16}$

Calculating the right-hand side:

$2 \div \frac{3}{4} = 2 \times \frac{4}{3}$ (dividing by a fraction is the same as multiplying by its reciprocal)

$= \frac{8}{3}$

$= 2\frac{2}{3}$

Therefore as $3\frac{1}{16} > 2\frac{2}{3}$

$\left(3\frac{1}{4} - 1\frac{1}{2}\right)^2 > 2 \div \frac{3}{4}$

Exercise 14.4

1. a. Explain why $20 \times 2\frac{5}{8} = \frac{20}{1} \times \frac{21}{8}$.
   b. Explain why $\frac{20}{1} \times \frac{21}{8} = \frac{5}{1} \times \frac{21}{2}$.
   c. Work out $20 \times 2\frac{5}{8}$.

2. a. Explain why $6\frac{1}{3} - 5\frac{4}{5} = 5\frac{4}{3} - 5\frac{4}{5}$.
   b. Explain why $5\frac{4}{3} - 5\frac{4}{5} = 5\frac{20}{15} - 5\frac{12}{15}$.
   c. Work out $6\frac{1}{3} - 5\frac{4}{5}$.

3. a. Explain how the diagram below shows that $4 \div \frac{1}{2} = 8$.

   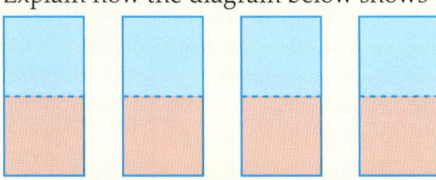

   b. Write $4 \div \frac{1}{2}$ as an equivalent multiplication.

4. Write down the reciprocal of the following values:
   a. 2
   b. 7
   c. $\frac{1}{2}$
   d. $\frac{1}{8}$
   e. $\frac{2}{3}$
   f. $\frac{3}{4}$
   g. Y
   h. $\frac{1}{X}$
   i. $\frac{X}{Y}$

5. Work out the following:
   a. $8 \times 5\frac{3}{4}$
   b. $7\frac{1}{3} \times 12$
   c. i) $15 \div \frac{1}{4}$
      ii) $15 \div \frac{3}{4}$
      iii) Comment on the relationship between your answers to (i) and (ii) above.
   d. i) $10 \div \frac{1}{5}$
      ii) $10 \div \frac{4}{5}$
      iii) Comment on the relationship between your answers to (i) and (ii) above.
   e. i) $4 \div \frac{1}{x}$
      ii) $4 \div \frac{3}{x}$
      iii) Comment on the relationship between your answers to (i) and (ii) above.

6. Calculate the following:
   a. $\left(\frac{2}{3} + 1\frac{4}{5}\right) \times 3$
   b. $\frac{2}{3} + 1\frac{4}{5} \times 3$
   c. $6\frac{3}{8} - 4\frac{1}{4} \div 2$
   d. $2\frac{5}{8} + \left(\frac{1}{4}\right)^2$

7. a. Which of the following fractions is bigger? Justify your answer.
      i) $\frac{2}{7}$ or $\frac{5}{7}$?
      ii) $\frac{5}{8}$ or $\frac{5}{11}$?

SECTION 2

b State whether the following statements are true or false.
  i) For fractions with the same denominator, the larger the numerator the larger the fraction.
  ii) For fractions with the same numerator, the larger the denominator the smaller the fraction.

8 a Copy the questions below. Insert the correct symbol (>, < or =) in the space provided.

  i) $4\frac{3}{10}$ .... $\frac{21}{5}$
  ii) $2\frac{9}{16}$ .... $\frac{41}{16}$
  iii) $5 - 2\frac{3}{5}$ .... $2\frac{1}{5}$
  iv) $\left(\frac{12}{5}\right)^2$ 6
  v) $\frac{17}{61}$ .... $\frac{27}{84}$
  vi) $4\frac{7}{43}$ .... $4\frac{2}{13}$

LET'S TALK
Parts v) and vi) are difficult to work out as fractions. Does working with a calculator and decimals make it easier? If so, why?

b Justify each of your answers in part (a).

 9 In the questions below, two fractions are given each time.
The one on the left is smaller than the one on the right.
Write down a fraction that is between the two fractions given.

  a $\frac{3}{14}$ .... $\frac{4}{14}$
  b $\frac{15}{26}$ .... $\frac{15}{25}$
  c $1\frac{5}{9}$ .... $1\frac{5}{6}$
  d $\frac{47}{8}$ .... $1\frac{9}{10}$
  e $\frac{6}{x}$ .... $\frac{7}{x}$

LET'S TALK
In parts (a)–(d), how would working with a calculator and decimals make the question easier?

Now you have completed Unit 14, you may like to try the Unit 14 online knowledge test if you are using the Boost eBook.

15 Manipulating algebraic expressions

- Understand how to manipulate algebraic expressions.

The language of algebra is often described as 'beautiful'. Its ability to simplify situations and provide clarity is one of its many beautiful features. Yet to fully appreciate it you have to learn the language. The more fluent you become in its rules and structures, the more its beauty reveals itself.

This unit will start to look at some of the ways in which algebraic expressions can be manipulated. These manipulations are central to becoming fluent in the language of algebra.

Expanding brackets

Consider the rectangle below with dimensions as shown.

To calculate the area involves multiplying the length by the width, i.e. $3(4a+5)$.

To see what the multiplication is equivalent to, the rectangle can be split into two parts with the area of each part as shown:

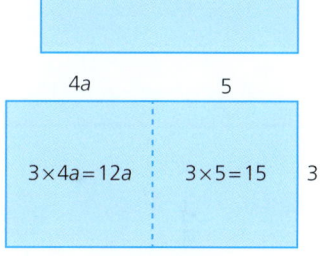

The area of the rectangle is therefore $12a + 15$.

But as the area is also $3(4a+5)$.

Then $3(4a+5) = 12a + 15$.

LET'S TALK
This is an example of *deductive logic*. What do you think *deductive logic* means in the context of this problem?

If the '3' outside the brackets multiplies both terms inside the brackets, then we get $12a + 15$.

In other words: $3 \times 4a + 3 \times 5 = 12a + 15$.

This is an example of the **distributive law** of multiplication, where the number outside the bracket multiplies each of the terms inside it.

The same method can be applied to more complex expressions.

SECTION 2

Worked example

a Write an expression for the area of the following rectangle.

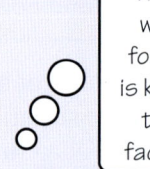

The expression written in the form $2a(5a+b)$ is known as giving the answer in factorised form.

The area is the multiplication of the length by the width: $2a(5a+b)$.

The expression can be given in expanded form by multiplying out the brackets:

$2a(5a+b) = 10a^2 + 2ab$

Multiplying $2a \times 5a$ can be considered as $2 \times 5 \times a \times a$

b Write an expression for the volume of the following cuboid.

The volume is the result of multiplying $(6a-1) \times 2a \times 3a$.

As multiplication is **commutative**, it can be done in any order:

$(6a-1) \times 2a \times 3a = (6a-1) \times 6a^2$

The volume of a cuboid is: length × width × height

Therefore, expanding the bracket gives:

$6a^2(6a-1) = 36a^3 - 6a^2$

An expression for the volume of the cuboid is $36a^3 - 6a^2$.

Exercise 15.1

1 Expand the following expressions.
  a $5a(2a+1)$
  b $4x(x+y)$
  c $2p(3p^2 - q)$
  d $3c^2(2d-3e)$
  e $\frac{1}{2}m^2(4m+6)$

2 Expand the following expressions and simplify your answer as much as possible.
  a $3(e+5)+2e-3$
  b $2(f-2)+3(f-1)$
  c $4f(2f-3)+5f(3f-2)$
  d $-2g(h+4)-3(h-2)$
  e $5h(2i+3)-7(i-1)$
  f $-3j(4+3j)-2(j-7)$

15 Manipulating algebraic expressions

 3 a Write an expression, using brackets, for the area of the triangle.

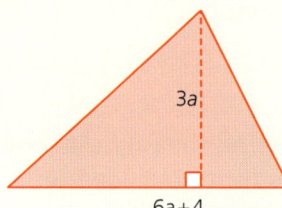

LET'S TALK
How many possible combinations of base length and perpendicular height could there be for a triangle with this area?

b The area of a different triangle is given as $A = 6a^2 + 12a$.
  i) Write a possible base length and a perpendicular height for the triangle.
  ii) Write another possible base length and perpendicular height for the triangle.

4 A sequence of three squares is shown opposite.
The first square has a side length of x cm. Each new square in the sequence has a side length double that of the previous one.
  a If the sequence is continued, how many squares are there if the total area is given by the expression $341x^2$ cm²?
  b The rule for the sequence is changed. The first shape is still a square with a side length of x cm. For subsequent shapes, the length is doubled as before, but the height increases by 1 cm as shown. Write an expression for the total area for the 4th pattern. Simplify your answer.

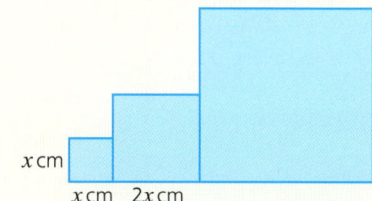

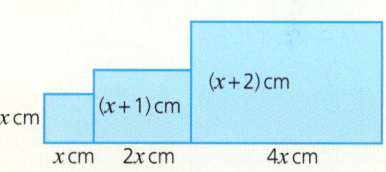

5 Two rectangles are shown below. The larger has dimensions $5x + 1$ cm and 10 cm, while the smaller has dimensions $2x - 1$ cm and 10 cm as shown:

6 cm of the smaller rectangle is placed over the larger rectangle so that only part of the larger rectangle is visible.
  a Write an expression for the area of the larger rectangle still visible.
  b The position of the smaller rectangle is moved. Write an expression for the smallest possible area of the larger rectangle that can be seen. Justify your answer.

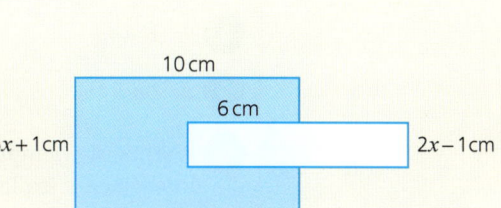

143

SECTION 2

> 2 is the highest common factor of both 6x and 8 so is written outside the bracket. Both the 6x and 8 are then divided by two to give 3x and 4. These are written in the brackets.

Expansion and factorisation

Changing $2(x+3)$ to $2x+6$ is called **expansion**.

The opposite of expansion, changing $2x+6$ to $2(x+3)$, is called **factorisation**.

To factorise an expression, we first need to find the highest common factor (HCF) of all the terms. We write this factor outside the brackets. We then divide each term by this factor to find what goes inside the brackets.

Worked example

a Factorise $6x+8$.
$6x+8 = 2(3x+4)$
The HCF of $6x$ and 8 is 2. So, this goes outside the brackets.

b Factorise $9a+6b+12$.
$9a+6b+12 = 3(3a+2b+4)$
The HCF of $9a$, $6b$ and 12 is 3. So, this goes outside the brackets.

Exercise 15.2

Factorise the following.

1. a $4a+10$ b $10a+15$ c $9a+21$
2. a $6b+3$ b $10b+5$ c $25b+10$
3. a $15c-25$ b $12c-8$ c $8a-24$
4. a $8-4d$ b $6-4d$ c $18-12d$
5. a $6a+4b$ b $7c+14d$ c $12a-16b$

 6. A student attempts to factorise the following expressions. In each case explain if she is right or wrong. If she is wrong, write down the correct answer.
  a $5-10b = 5(-2b)$
  b $24b-12a = 6(4b-2a)$
  c $27a+27b = 27(a+b)$

15 Manipulating algebraic expressions

Sometimes, the highest common factor is a letter rather than a number.

> *b* is the only factor of all three terms so is written outside the brackets.
>
> The three terms are then divided by *b* to produce $2a + 3c + 4d$ inside the brackets.

Worked example

a Factorise $2ab + 3bc + 4bd$.
   $2ab + 3bc + 4bd = b(2a + 3c + 4d)$
   The HCF is *b*, so this goes outside the brackets.

b Factorise $ax + bx + x^2$.
   $ax + bx + x^2 = x(a + b + x)$
   The HCF is *x*, so this goes outside the brackets.

Exercise 15.3

Factorise the following.

1. a $2ax + 3bx + 4cx$ b $7ab - 8bc$
2. a $3pq - 4q + 5qs$ b $2mn - 3nr + 5np$
3. a $4ax + 3x^2$ b $4ab - 3b^2$
4. a $6p^2 - 5pq$ b $7mn - 2m^2$
5. a $x^2 - ax$ b $pqr - p^2$

Sometimes, the highest common factor is a combination of numbers and letters.

Worked example

a Factorise $8x^2 - 4ax + 6bx$.
   $8x^2 - 4ax + 6bx = 2x(4x - 2a + 3b)$
   The HCF is $2x$, so this goes outside the brackets.

b Factorise $30abc - 24abd - 12a^2 b$.
   $30abc - 24abd - 12a^2 2b = 6ab(5c - 4d - 2a)$
   The HCF is $6ab$, so this goes outside the brackets.

SECTION 2

Exercise 15.4

Factorise the following expressions fully.

1. a $4xy - 6yz$
   b $9pq - 12qr$

2. a $15mn - 10pm$
   b $14bc - 21c^2$

3. a $6pq - 30p^2$
   b $15x^2 - 10xy$

4. a $12x^2y - 8xy^2$
   b $10ab^2 - 25a^2b$

5. a $7ax - 14ay + 21az$
   b $30ax^2 - 6bx^2 + 9cx^2$

Exercise 15.5

1. a Which of the following expressions are not fully factorised?

   i) $6a(5a + b)$
   ii) $x(7x + 14)$
   iii) $2pq(pq + 2q^2)$
   iv) $2mn(6m - 4n^2)$
   v) $8cd(6c + 9cd)$
   vi) $7m^2n^2(12m - 3m^2n)$

   b For each of the ones you identified in part (a) write them in fully factorised form.

2. The area of this triangle is given by the expression $6x + 10$.

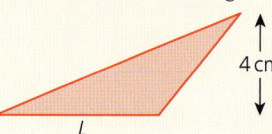

Write an expression for the base length L, giving your answer in terms of x.

3. Explain why the circumference of the circle with a diameter of $14x - 10$ as shown can be given by the expression $2\pi(7x - 5)$.

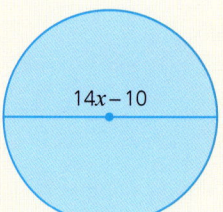

15 Manipulating algebraic expressions

4 Show that the area of the cross is given by the expression $12(6x+1)$.

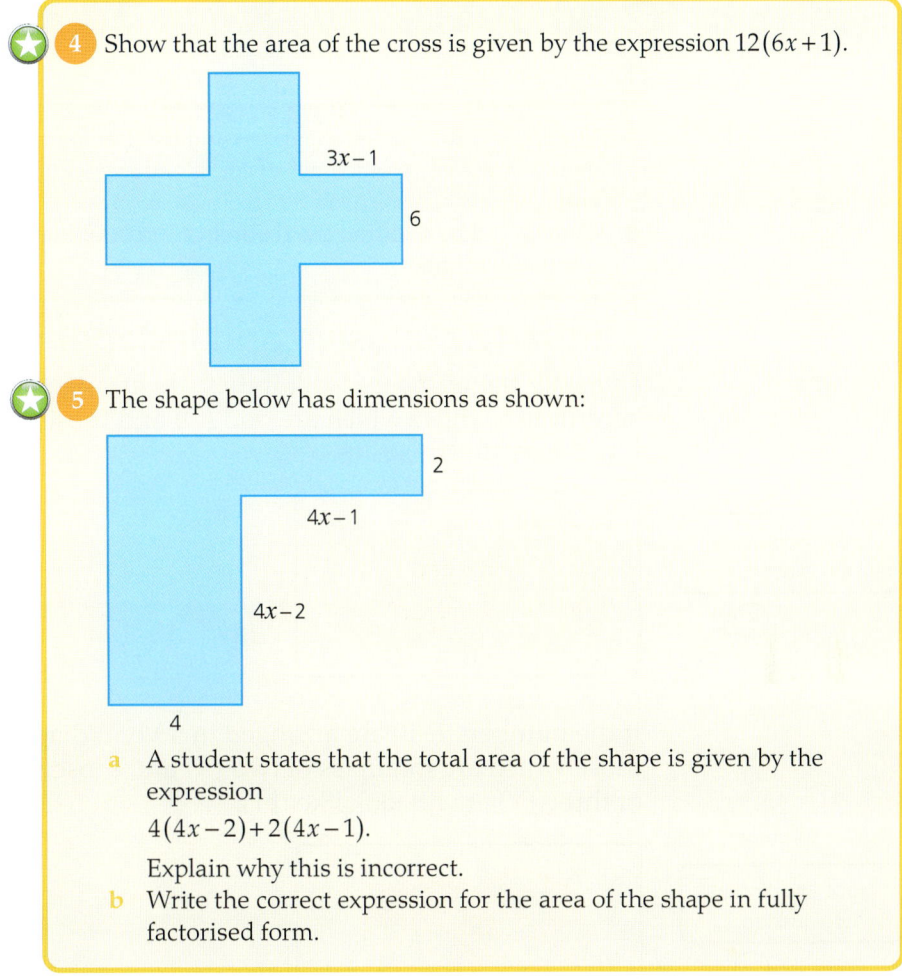

5 The shape below has dimensions as shown:

   a A student states that the total area of the shape is given by the expression
      $4(4x-2)+2(4x-1)$.
      Explain why this is incorrect.
   b Write the correct expression for the area of the shape in fully factorised form.

Now you have completed Unit 15, you may like to try the Unit 15 online knowledge test if you are using the Boost eBook.

16 Combined events

- Understand that tables, diagrams and lists can be used to identify all mutually exclusive outcomes of combined events (independent events only).
- Understand how to find the theoretical probabilities of equally likely combined events.

Mutually exclusive and independent events

Consider the numbers 1–10. If they are arranged in a Venn diagram according to those which are odd (O) and those which are even (E), it will look like this:

Here the circles do not overlap.

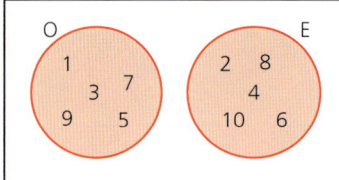

If the numbers 1–10 are arranged in a Venn diagram according to those which are multiples of two (X) and those which are multiples of three (Y), it will look like this:

Here the circles do overlap.

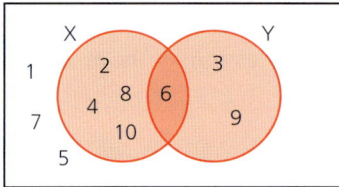

LET'S TALK
Can you think of another way of dividing the ten numbers into mutually exclusive outcomes?

In the first case, the circles of the Venn diagram do not overlap as there is no number which can belong to both groups. i.e. no number can be both odd and even. These outcomes are said to be **mutually exclusive**.

If a spinner is spun and a dice is rolled, the outcome of the dice has no effect on the outcome of the spinner. Outcomes which have no effect on each other are said to be **independent events**.

When working with the probability of combined events, several strategies for visualising the outcomes can be used.

16 Combined events

Worked example

Two spinners are shown below:

Spinner 1 has four **equally likely** outcomes, Blue (B), Green (G), Orange (O) and Yellow (Y), while spinner 2 has three equally likely outcomes, Yellow (Y), Pink (P) and Blue (B).

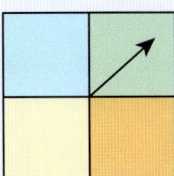 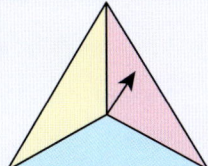

i) Both spinners are spun. How many possible outcomes are there?

Spinner 1	Spinner 2
Blue	Yellow
Blue	Pink
Blue	Blue
Green	Yellow
Green	Pink
Green	Blue
Orange	Yellow
Orange	Pink
Orange	Blue
Yellow	Yellow
Yellow	Pink
Yellow	Blue

Notice how the list has been arranged in a logical order.

LET'S TALK
Why is a logical order helpful?
Can you think of another logical order to identify all 12 possible outcomes?

LET'S TALK
What are the advantages and disadvantages of a list for identifying all the possible outcomes?

From the ordered list we can see that there are 12 possible outcomes.

ii) What is the probability of both spinners landing on yellow?
As there are 12 equally likely outcomes and only one shows both spinners landing on yellow, then $P(YY) = \frac{1}{12}$.

iii) What is the probability of both spinners showing the same colour?
This implies either both landing on blue or both landing on yellow.
$P(BB) + P(YY) = \frac{1}{12} + \frac{1}{12} = \frac{2}{12} = \frac{1}{6}$

SECTION 2

The previous page example was solved by producing a logical list of all the possible outcomes. However, there are other possible diagrams which can be used to solve the same problem including **sample space diagrams** and **tree diagrams**.

A sample space diagram is similar to a two-way table in that all the outcomes are presented in a grid like formation as shown below:

> **LET'S TALK**
> How does this sample space diagram show how many outcomes involve either spinner landing on blue? What about only one spinner landing on blue?
>
> What are the advantages and disadvantages of a sample space diagram?

		Spinner 1			
		Blue	Green	Orange	Yellow
Spinner 2	Yellow	BY	GY	OY	YY
	Pink	BP	GP	OP	YP
	Blue	BB	GB	OB	YB

The sample space diagram shows that there are 12 possible outcomes and it is easy to see that only one outcome shows both spinners landing on yellow (YY).

A tree diagram, as its name implies, looks like a tree (usually lying on its side).

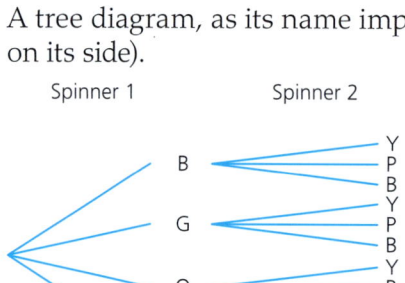

Notice how all the outcomes of spinner 1 are listed on the left, then off each of those outcomes, the three possible outcomes of spinner 2 are given, leading to 12 possible outcomes in total.

To use the tree diagram to calculate the probability of both spinners landing on yellow, those outcomes are highlighted as shown:

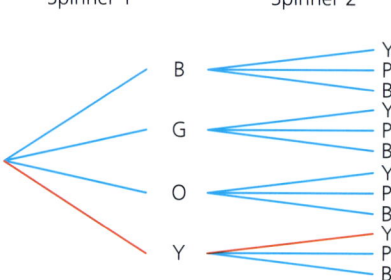

> **LET'S TALK**
> What are the advantages and disadvantages of using a tree diagram?

16 Combined events

This is only one combination out of a possible 12. Therefore,

$$P(YY) = \frac{1}{12}$$

Exercise 16.1

 1 a A four-sided dice, numbered 1–4, is rolled and an ordinary coin is flipped.
   i) Construct a sample space diagram to show all the possible outcomes.
   ii) How many possible outcomes are there?
   iii) What is the probability of getting a head on the coin and a 2 on the dice, i.e. P(H2)?
   iv) What is the probability of not getting a head on the coin or a 2 on the dice?
   b Repeat the questions above but using a tree diagram. Show your method clearly for parts (iii) and (iv).
   c Repeat the questions in part (a) above using an ordered list. Show your method clearly for the answers to parts (iii) and (iv).
   d **Critique** the three methods. Which method did you find easiest to use?

 2 a Two four-sided dice, each numbered 1–4, are rolled. The scores on the two dice are added together.
   i) Copy and complete the sample space diagram to show the possible outcomes.

		Dice A			
		1	2	3	4
Dice B	1				
	2		4		
	3				7
	4				

   ii) How many possible outcomes are there?
   iii) What is the probability of getting a total score of 6?
   iv) What is the probability of not getting a total score of 6?
   v) How many times more likely is a total score of 5 than a total score of 3?
   b Explain why a tree diagram is likely to be more complicated when answering the questions in part (a). You do not need to draw the tree diagram.
   c Explain why an ordered list is likely to be more complicated that the space diagram in answering the questions in part (a).

SECTION 2

3 An ice cream seller sells three flavours of ice cream. These are vanilla, chocolate and caramel flavour. In addition to this, customers can choose one of two toppings, either chocolate sauce or sprinkles, or choose no topping.
  a He decides to write down all the possible options and starts the list as follows:
   - Vanilla + no topping
   - Sprinkles + caramel
   - Chocolate sauce and vanilla

   Comment on whether this is a good way of starting the list.
  b Write an ordered list of all the possible combinations of ice creams and toppings.
  c Assuming that each option is equally likely calculate:
   i) the probability that someone will choose a vanilla ice cream with chocolate sauce
   ii) the probability that someone will have no topping.

4 Two ordinary six-sided dice, each numbered 1–6, are rolled. The scores on the two dice are added together.
  a Copy and complete the sample space diagram to show the possible outcomes.

		Dice A					
		1	2	3	4	5	6
Dice B	1						
	2		4				
	3				7		
	4						
	5						
	6						

  b How many possible outcomes are there?
  c i) Which total score is the most likely outcome?
     ii) What is the probability of getting the most likely outcome?
     iii) What is the probability of not getting the most likely outcome?
  d What is the probability of getting a total that is an even number?
  e How many times more likely is a total score of 6 than a total score of 3?

5 Four unbiased coins are flipped. Three show heads and one shows tails. A student wants to find out the probability of this happening.
  a Explain why a space diagram will not help to answer this question.
  b Using either a tree diagram or an ordered list, calculate the probability of getting three heads and one tail.

> **LET'S TALK**
> Is the probability of getting three heads and one tail in the order H,H,H,T as likely as getting them in any order?
>
> If you flipped six coins, how would you calculate the probability of getting two heads and four tails in *any* order?

 Now you have completed Unit 16, you may like to try the Unit 16 online knowledge test if you are using the Boost eBook.

152

17 Constructions, lines and angles

- Construct triangles, midpoint and perpendicular bisector of a line segment, and the bisector of an angle.
- Derive and use the fact that the exterior angle of a triangle is equal to the sum of the two interior opposite angles.
- Recognise and describe the properties of angles on parallel and intersecting lines, using geometric vocabulary such as alternate, corresponding and vertically opposite.

> A pair of compasses means one of them, not two. The term has the same meaning as a pair of scissors or a pair of trousers.

Constructing triangles

A triangle can be constructed in various ways depending on the information given.

In some cases, it can be done with just a protractor and ruler, but in other cases it will need the use of a pair of compasses.

This section will cover the four main types of triangle construction. These are identified in the points and diagrams that follow.

- When two angles and the included side are given:

> When two angles and the included side are given this is known as an ASA type triangle.

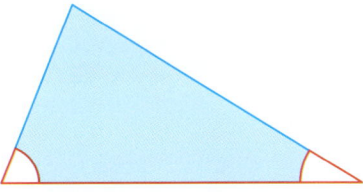

- When two sides and the included angle are given:

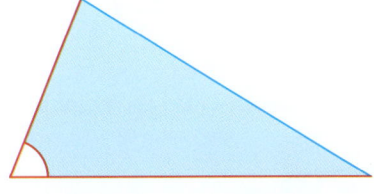

> When two sides and the included angle are given this is known as an SAS type triangle.

- When the three sides are given:

> When the three sides are given this is known as an SSS type triangle.

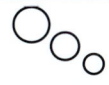

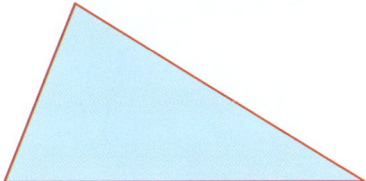

153

SECTION 2

> A right-angled triangle where the hypotenuse and one other side are given is known as a **RHS** type triangle.

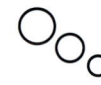

- When we are given a right angle, the **hypotenuse** and another side:

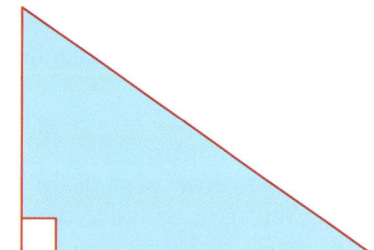

Worked example

Draw a triangle ABC where the length of side AB is 8.2 cm, ∠A = 52° and ∠B = 40°.

Use a ruler and protractor.
- First draw a line 8.2 cm long and label its ends A and B.

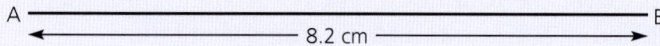
8.2 cm

 Make sure to leave enough room above the line for the rest of the triangle.

- Then use your protractor as shown to mark angles A and B.

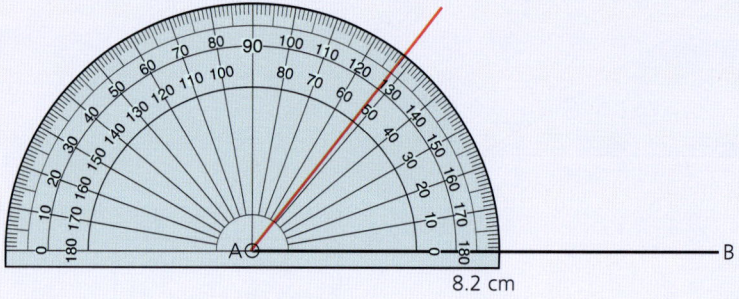
8.2 cm

LET'S TALK
How do we know which scale of the protractor to use?

- Extend each of the lines drawn so that they intersect. Label the point of intersection C.

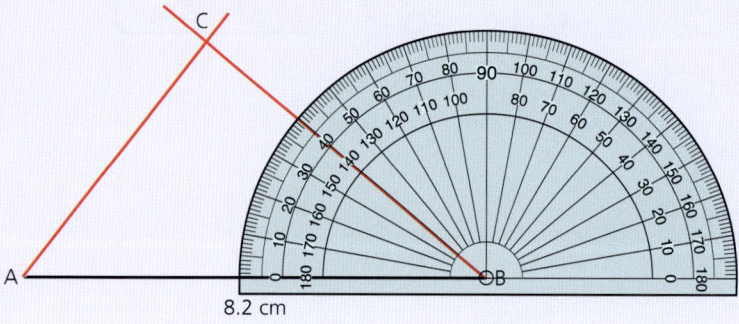
8.2 cm

17 Constructions, lines and angles

Exercise 17.1

Construct the triangles ABC in questions 1–5. After each construction, measure the sides AC and BC and angle C. You will need a ruler and a protractor.

1. AB = 7 cm, ∠A = 50°, ∠B = 40°.
2. AB = 8.5 cm, ∠A = 30°, ∠B = 60°.
3. AB = 9.0 cm, ∠A = 45°, ∠B = 45°.
4. AB = 8.3 cm, ∠A = 100°, ∠B = 30°.
5. AB = 7.6 cm, ∠A = 112°, ∠B = 41°.
6. Explain why the following triangle ABC cannot be constructed using the method above for ASA type triangles.
   AB = 5.2 cm, ∠A = 49°, ∠C = 32°.
7. Without constructing it, explain whether the following triangle XYZ can be constructed.
   XY = 12.5 cm, ∠X = 98°, ∠Y = 85°.

Worked example

Construct the triangle ABC, where the length AB = 8 cm, AC = 7 cm and ∠A = 60°.

You will need a ruler and a protractor.
- First draw a line 8 cm long and label its ends A and B.
- Then use your protractor to mark an angle of 60° at A. Use your ruler to draw a straight line from A to the mark and extend the line.

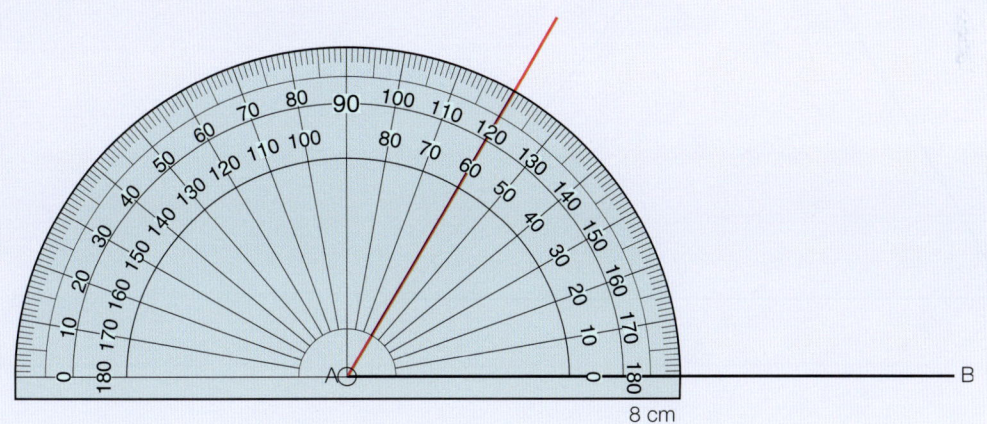

SECTION 2

- Measure 7 cm from A along this line. Label this point C.

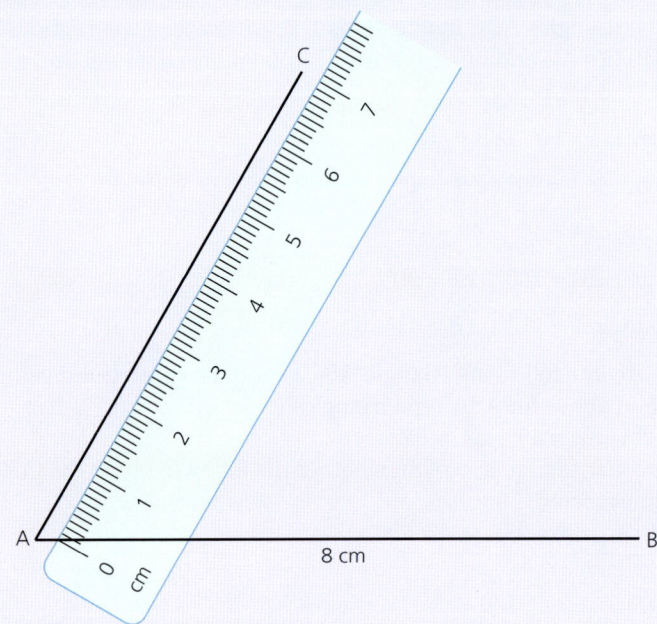

- Now draw a straight line from B to C. This completes the triangle ABC.

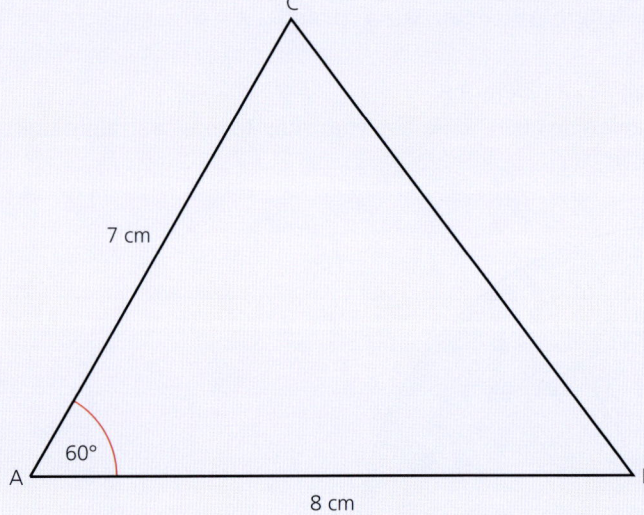

17 Constructions, lines and angles

Exercise 17.2

Construct the following triangles ABC.

1. AB = 10 cm, AC = 8 cm, ∠A = 45°.
2. AB = 13 cm, AC = 12 cm, ∠A = 35°.
3. AB = 8 cm, AC = 10 cm, ∠A = 100°.
4. AB = 6.8 cm, AC = 6.8 cm, ∠A = 72°.

LET'S TALK
What is meant by the 'included angle' in ASA-type triangles?

Worked example

Construct the triangle XYZ, where the length XY = 10 cm, XZ = 8 cm and YZ = 6 cm. Use a ruler and a pair of compasses only.

- First use a ruler to draw a line 10 cm long and label its ends X and Y.
- Open your compasses to 8 cm. Place the compass point on X and draw an arc.

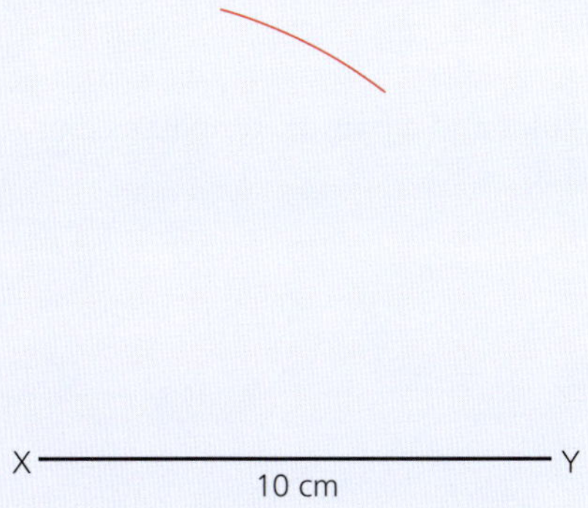

Make sure that your pair of compasses have been tightened.

X ————————— Y
       10 cm

LET'S TALK
What can be said about the distance of every point on the arc from X?

- Open your compasses to 6 cm. Place the compass point on Y and draw another arc. Make sure that the two arcs intersect.

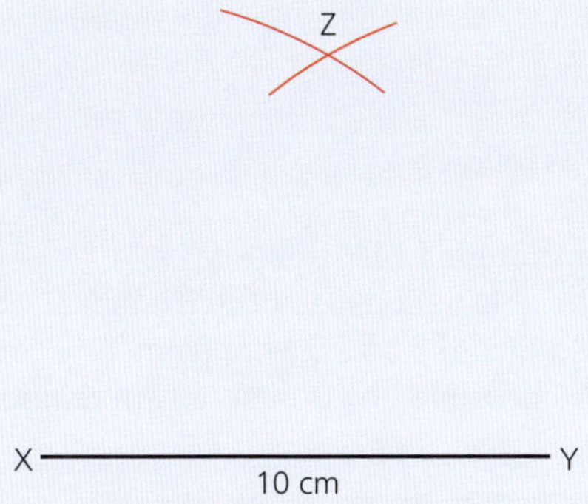

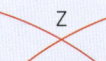

X ————————— Y
       10 cm

LET'S TALK
What can be said about the distance of every point on the second arc from Y?

SECTION 2

The only point that is 8 cm from X and 6 cm from Y is where the two arcs intersect. Label this point Z.
- Finally, draw straight lines from X to Z and from Y to Z. This completes the triangle XYZ.

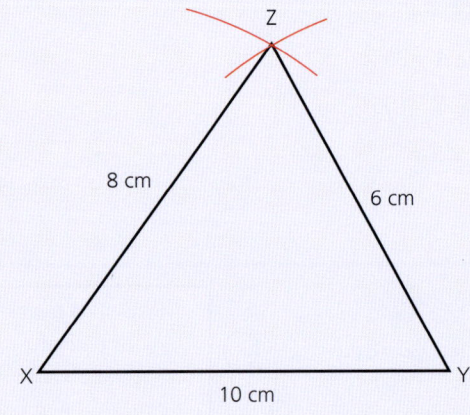

Exercise 17.3

For questions 1–4, construct the triangles PQR using only a ruler and a pair of compasses. For each triangle, measure accurately each of the angles P, Q and R.

1. PQ=6 cm, PR=8 cm, QR=10 cm
2. PQ=12 cm, PR=9 cm, QR=5 cm
3. PQ=4 cm, PR=7 cm, QR=7 cm
4. PQ=6.2 cm, PR=6.2 cm, QR=6.2 cm
5. Name the type of triangle constructed in each of questions 1–4.
6. a Without constructing them, deduce which of the three triangles PQR below cannot be constructed. Justify your choice.
    i) PQ=12 cm, PR=5 cm, QR=6 cm
    ii) PQ=14 cm, PR=8 cm, QR=10 cm
    iii) PQ=7 cm, PR=9 cm, QR=14 cm
   b The lengths of two sides of a triangle LMN are given as follows:
   LM=20 cm, LN=8 cm.
    i) Write a value for the length of side MN so that the triangle LMN is isosceles.
    ii) What length must MN be bigger than in order to be able to construct the triangle?
7. Two boats, S and T, are 8.5 km apart. A third boat, U, is 6 km from S and 5.5 km from T. Construct a diagram to a scale of 1 cm to 1 km, to show the possible position(s) of U in relation to both S and T.

17 Constructions, lines and angles

Right-angled triangles

In a right-angled triangle, two of the sides are perpendicular (at right angles) to each other. The third side (the one that is opposite the right angle) is the longest side and is called the hypotenuse.

A right-angled triangle can be constructed using a ruler and a pair of compasses.

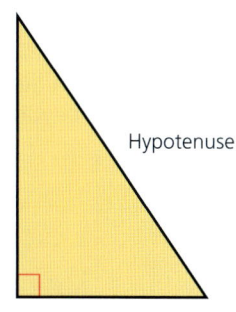
Hypotenuse

Right angle (90°)

Worked example

Construct the right-angled triangle ABC, where the length AB = 6 cm and the hypotenuse BC = 11 cm.

- Using a ruler, draw a line longer than 6 cm. Mark off a 6 cm length and label it AB.

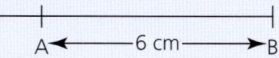

- Open the pair of compasses to about 2 cm. Place the compass point on A and mark two arcs on the line, one either side of point A.

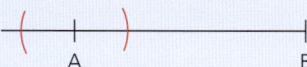

- Open the compasses further, to about 4 cm. Place the compass point on the intersection of the line and one of the small arcs and draw another arc above A.

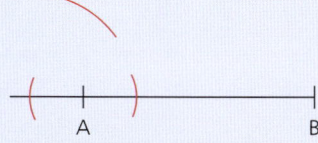

- Keeping the same radius, place the point on the intersection of the line with the other small arc and draw another arc above A. Make sure that it intersects the first one.

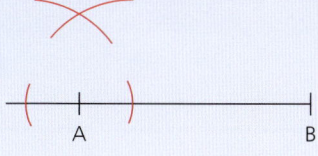

SECTION 2

- Using a ruler, draw a line from point A through the intersection of the two arcs. This line is perpendicular to AB, so the point C will be on it. As we are not told the length of AC, extend the line beyond the intersection of the two arcs.

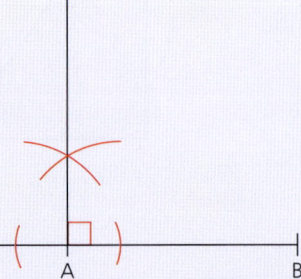

- The hypotenuse is the side opposite the right angle. Open the compasses to 11 cm. Place the compass point on B and draw an arc so that it intersects the line which is perpendicular to AB.

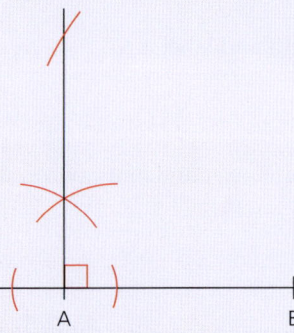

- Label the point of intersection C. Use a ruler to draw the line BC. This completes the triangle ABC.

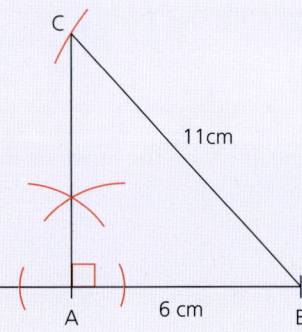

17 Constructions, lines and angles

Exercise 17.4

Construct the following right-angled triangles using only a ruler and a pair of compasses. For each triangle, measure accurately the length of the third side.

1
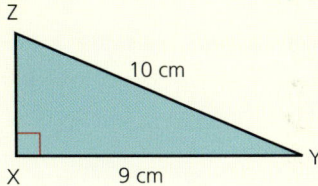

2
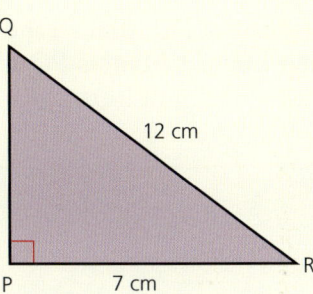

3
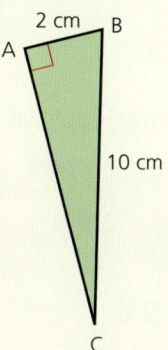

4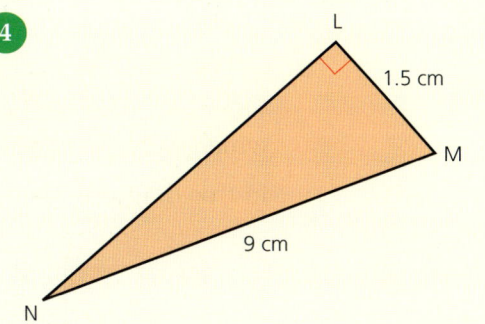

5 In the triangle DEF, the length DX=4 cm, XE=6 cm and EF=8 cm.

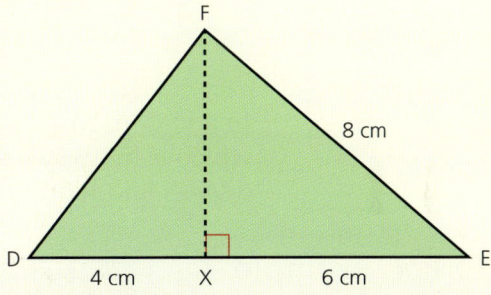

a Using a pair of compasses and a ruler only, construct the triangle DEF.
b Measure the length of the side DF.
c Measure the height of the triangle, XF.

6 In the obtuse-angled triangle PQR, PX and XR are perpendicular to each other. The length XQ=5 cm, PQ=10 cm and PR=15 cm.

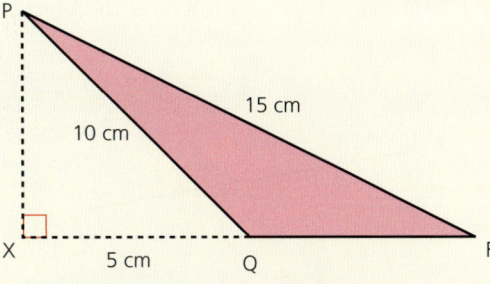

a Construct the triangle PQR using only a ruler and a pair of compasses.
b Measure the length of QR.
c Measure the size of the angle PQR.

SECTION 2

The perpendicular bisector and the midpoint

Lines that are at right angles to each other are said to be **perpendicular**.

The right-angled triangles in the previous section had two sides which were perpendicular to each other.

In this diagram, the lines PQ and ST are perpendicular to each other.

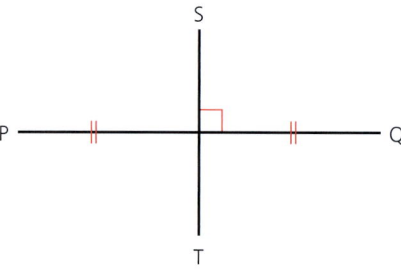

In addition, ST passes through the **midpoint** of PQ, that is, the point which is halfway between P and Q. Therefore, ST **bisects** PQ (it divides it in half). ST is the perpendicular bisector of PQ.

> **Worked example**
>
> Construct the perpendicular bisector of a line using a ruler and a pair of compasses.
> - Draw a line and label it AB.
>
>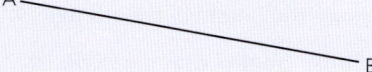
>
> - Open a pair of compasses to more than half the distance AB.
> - Place the compass point on A and draw arcs above and below AB.
> - Keeping the same radius, place the compass point on B and draw arcs above and below AB. Make sure that they intersect the first pair of arcs.
>
>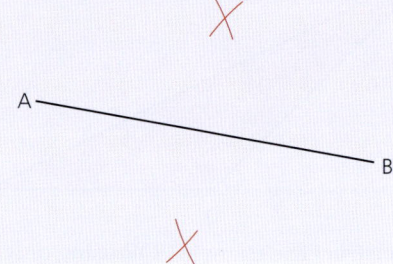

162

17 Constructions, lines and angles

- Draw a line through the two points where the arcs intersect.

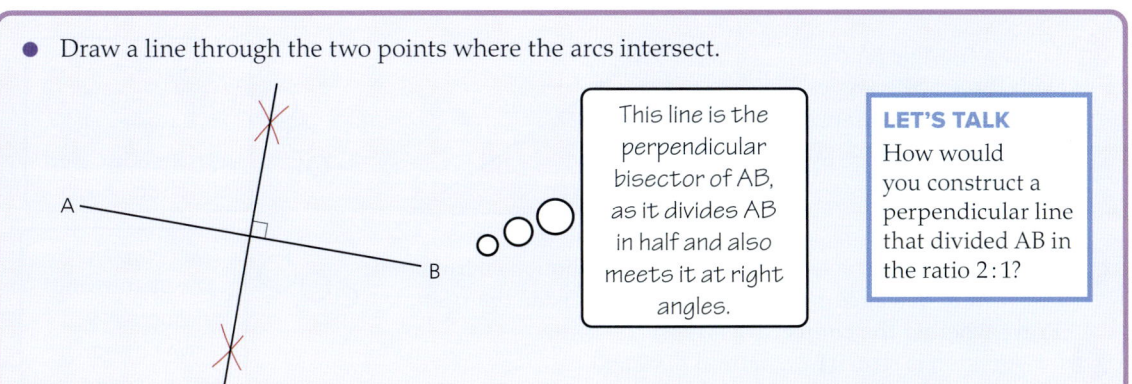

This line is the perpendicular bisector of AB, as it divides AB in half and also meets it at right angles.

LET'S TALK
How would you construct a perpendicular line that divided AB in the ratio 2:1?

Exercise 17.5

1. Copy each of the lines drawn below (actual size) onto plain paper and construct their perpendicular bisectors.

   a   7 cm

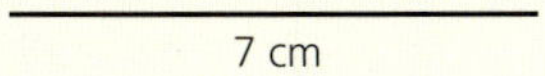

   b   11 cm

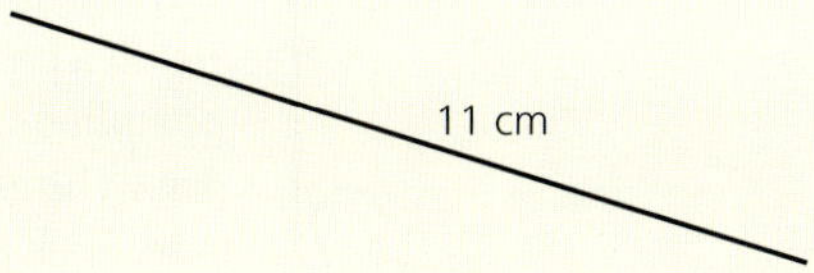

   c   8 cm

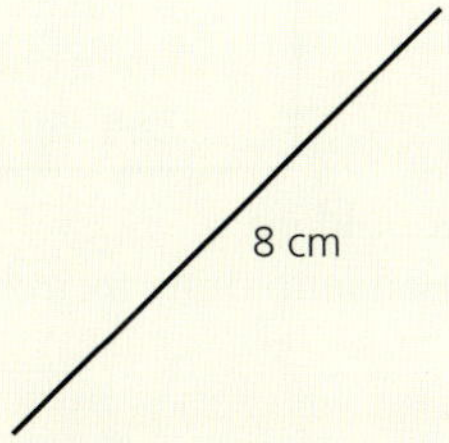

   d   9 cm

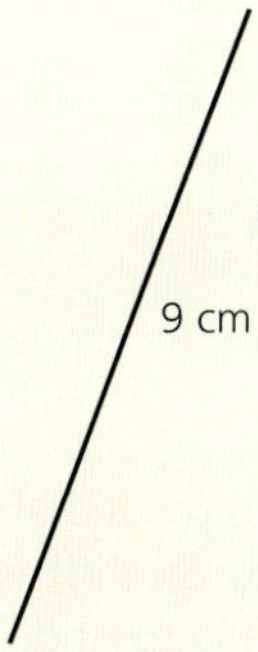

   When constructing a perpendicular bisector, it is important to leave the construction arcs in the final diagram.

2. Copy this diagram.
   a   Construct the perpendicular bisector of AB.
   b   Construct the perpendicular bisector of BC.
   c   What can be said about the point of intersection of the two perpendicular bisectors?
   d   Construct the perpendicular bisector of AC. What do you notice?
   e   Is it possible to draw a circle so that points A, B and C all lie on the circumference of the circle? Give a **convincing** reason for your answer.

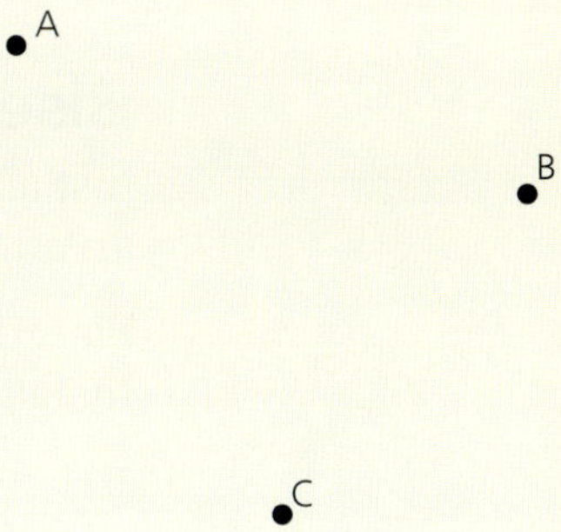

SECTION 2

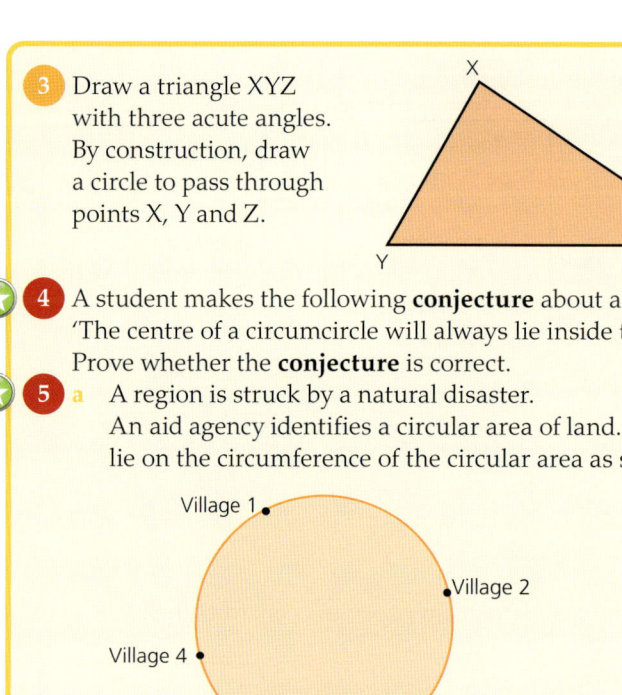

3 Draw a triangle XYZ with three acute angles. By construction, draw a circle to pass through points X, Y and Z.

> A circle which passes through all three vertices of a triangle is called the **circumcircle**.

4 A student makes the following **conjecture** about a circumcircle:
'The centre of a circumcircle will always lie inside the triangle.'
Prove whether the **conjecture** is correct.

> To prove that the statement is correct you have to show it is the case for *all* triangles. To prove that it is incorrect you only have to show it is not the case for one triangle.

5 a A region is struck by a natural disaster.
An aid agency identifies a circular area of land. Four villages lie on the circumference of the circular area as shown.

> Draw around a circular object rather than using a pair of compasses to draw the initial circle. Then check your answer using a pair of compasses.

The aid agency wants to put a distribution centre that is **equidistant** from all four villages. By **modelling** the situation, show that, by construction, it is possible to find where the distribution centre should go.

b Assuming four points do not lie on a straight line, is it always possible to draw a circle that passes through all four points? Prove your answer.

Bisecting an angle

Just as bisecting a line means dividing it into two halves, bisecting an angle means splitting the angle in half.

Bisecting an angle properly does not involve using a protractor. Instead, a pair of compasses should be used.

17 Constructions, lines and angles

Worked example

Bisect the angle A using a pair of compasses.

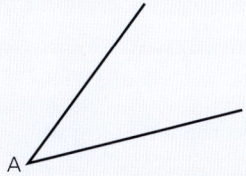

- Open the pair of compasses. Place the compass point on A and draw two small arcs so that they intersect each of the arms of the angle.

- Place the compass point on each of the points where the arcs intersect the lines and draw two more arcs that intersect. Make sure the radius remains unchanged.

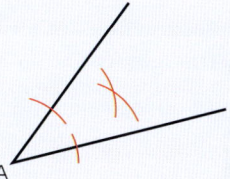

- With a ruler, draw a straight line from A through the point of intersection of these two arcs. This line bisects angle A.

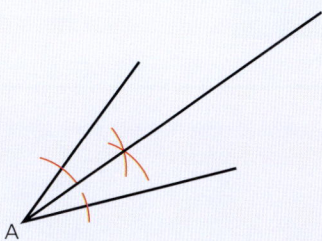

Exercise 17.6

For questions 1–4, draw angles similar to the ones shown. Bisect each of them using only a ruler and a pair of compasses.

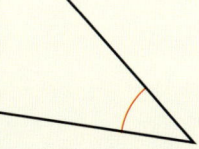

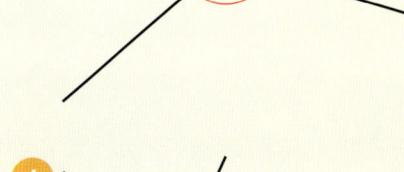

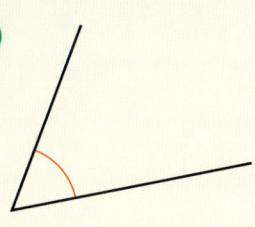

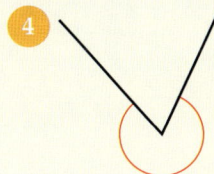

SECTION 2

5. a Using only a ruler and a pair of compasses, construct two lines which are perpendicular to each other.
   b Bisect the right angle. What size angle has been constructed?
6. a Construct an equilateral triangle ABC of side length 6 cm.
   b i) Bisect angle A.
      ii) Extend the bisecting line until it intersects the side BC. Label this point X.
      iii) Describe the triangle ABX.

For questions 7 and 8, use only a ruler and a pair of compasses to construct the following shapes.

7.

8.

9. A student states that 'as it is possible to construct an equilateral triangle, then it is possible to construct an angle of 15°'. Prove that this statement is correct.

LET'S TALK
How is it possible to use an equilateral triangle to construct an angle of 45°?

17 Constructions, lines and angles

Angle properties and proof; Alternate and corresponding angles

In Stage 7, you investigated the relationships between angles at a point on a straight line, around a point, and between intersecting lines.

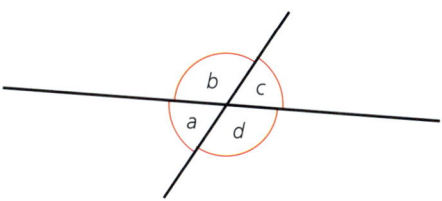

You saw that angles at a point on a straight line add up to 180°. In the diagram above, for example, $a + b = 180°$. They are called **supplementary angles**.

> **LET'S TALK**
> Can you remember how to **prove** that $a = c$?

You also saw that angles around a point add up to 360°, for example $a + b + c + d = 360°$, and that **vertically opposite angles** are equal in size, for example $a = c$.

In addition, you were introduced to two further relationships between the angles formed within **parallel** and intersecting lines.

In the diagram below, angles a and g are equal. They are called **alternate angles**. Alternate angles can be found by looking for a '**Z**' formation in a diagram.

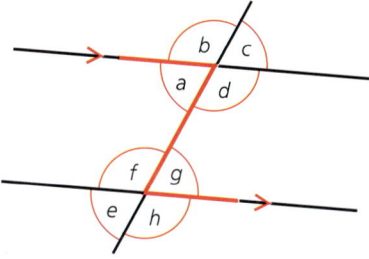

> The F formation can also be backwards and/or upside down.

Remember that the Z formation can be backwards too.

In the diagram on the next page, angles a and e are equal. They are called **corresponding angles**. Corresponding angles can be found by looking for an '**F**' formation within a diagram.

167

SECTION 2

The angles b and f are another pair of corresponding angles, so $b = f$.

We know that $a + b = 180°$, so it follows that $a + f = 180°$ and that angles a and f are supplementary.

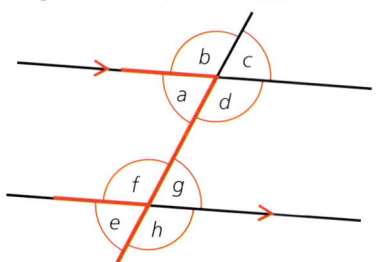

Exercise 17.7

> You need to use your **specialising** skills to find suitable pairs of angles.

Use this diagram for questions 1–6.

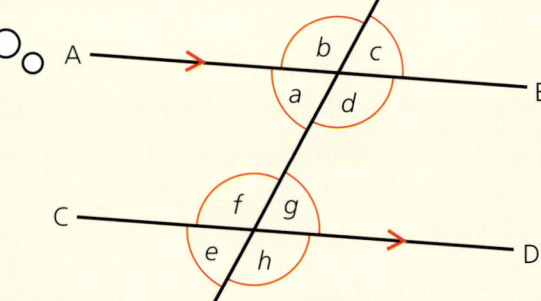

1. Write down as many pairs of angles as you can which add up to 180° because they are angles on a straight line.
2. Write down as many groups of angles as you can which add up to 360° because they are angles around a point.
3. Write down as many pairs of supplementary angles as you can.
4. Write down as many pairs of vertically opposite angles as you can.
5. Write down as many pairs of alternate angles as you can.
6. Write down as many pairs of corresponding angles as you can.

17 Constructions, lines and angles

Exercise 17.8

Calculate the size of each of the unknown angles in the diagrams in questions 1–8.
Justify your answers using the correct mathematical language.

1

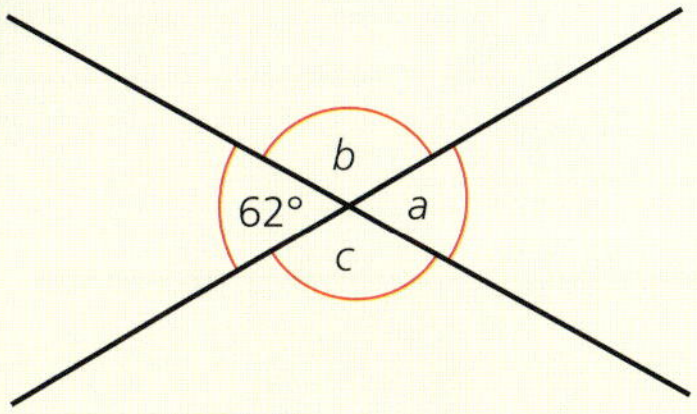

2

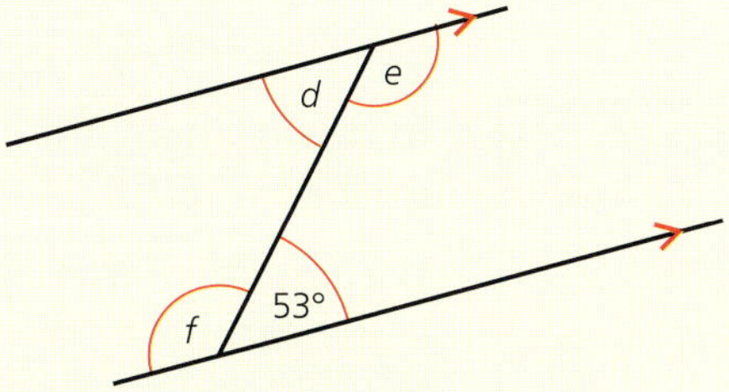

3

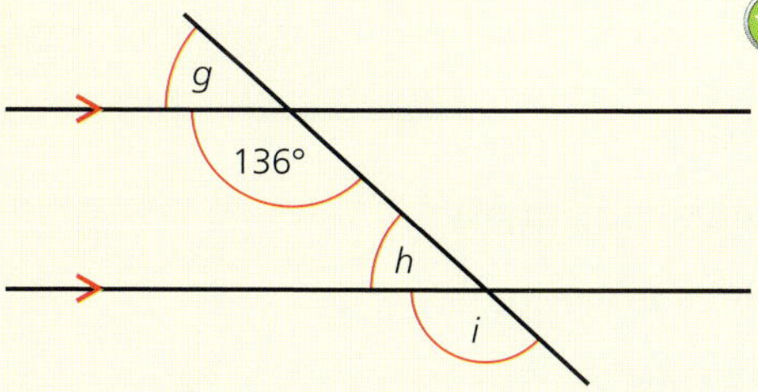

4

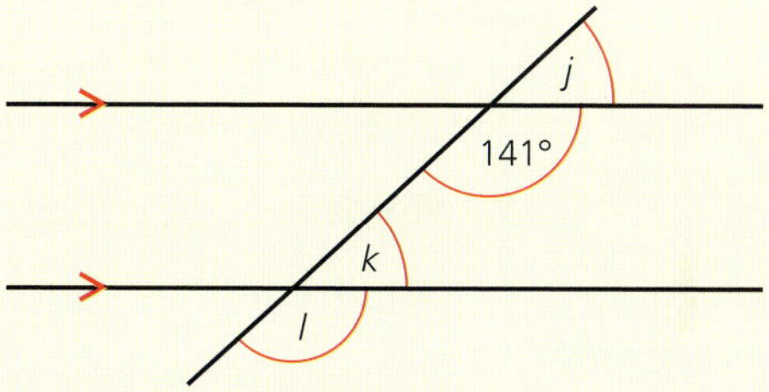

5

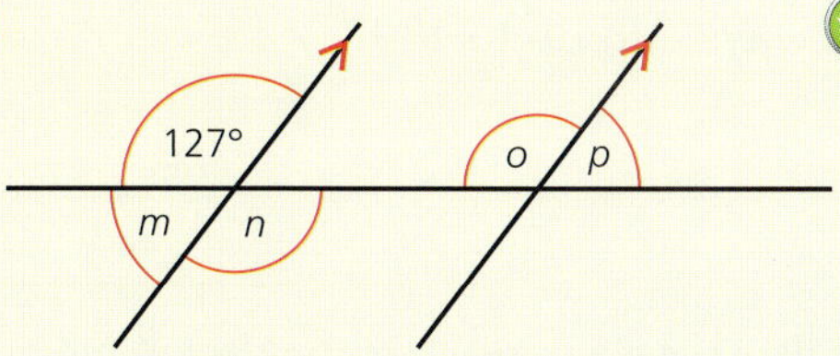

6

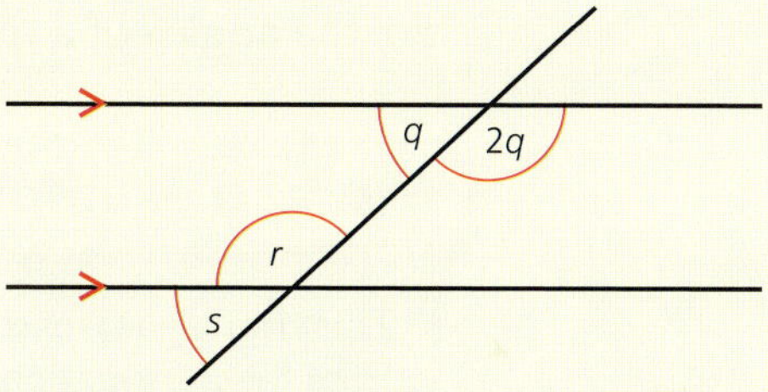

7

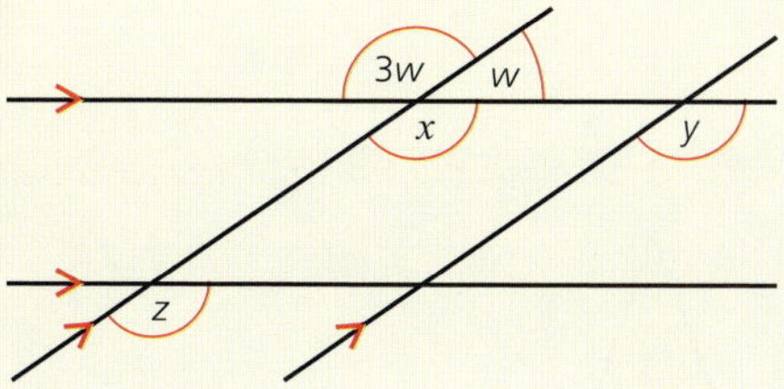

8

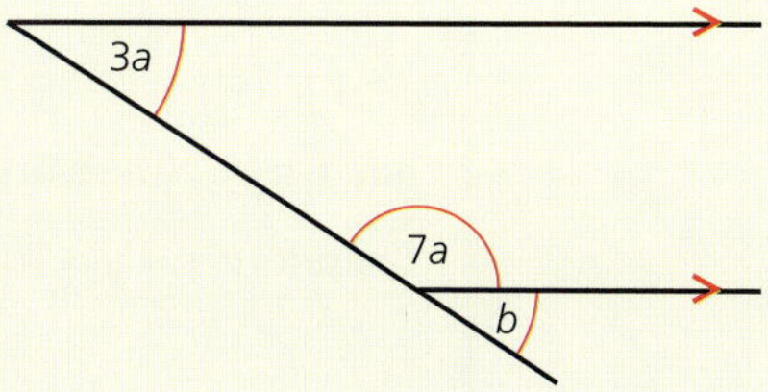

SECTION 2

Angles and triangles

A triangle has three angles as shown.

Although the sizes of the individual angles can vary, their total is always constant.

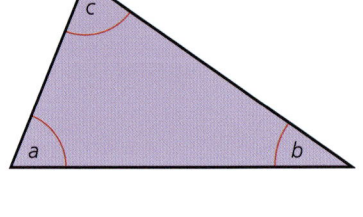

The three angles always add up to 180°. In this triangle, for example, $a + b + c = 180°$.

We can prove this by drawing a line parallel to the base of the triangle and passing through its apex (highest point).

$c + d + e = 180°$ (angles on a straight line add up to 180°)

But $a = d$ (alternate angles are equal)

and $b = e$ (alternate angles are equal)

Therefore, $c + d + e = a + b + c = 180°$.

So, the angles of a triangle always add up to 180°.

> **LET'S TALK**
> The three angles of a triangle drawn on flat paper add up to 180°. Do you think the three angles of a triangle drawn on a sphere follow the same rule?
>
> Use the internet to research this.

Worked example

This triangle is isosceles. Calculate the size of each of the base angles from the information given.

$x + y + 40° = 180°$ (angles of a triangle add up to 180°)

$x + y = 180° - 40° = 140°$

But $x = y$ (base angles of an isosceles triangle are equal)

Therefore $x = y = 70°$.

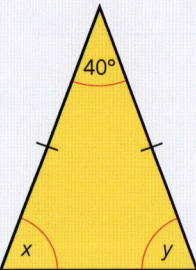

The **exterior angle** of a triangle is also related to the interior angles. An exterior angle is found by extending one of the sides of the triangle. In this diagram, angle d is an exterior angle.

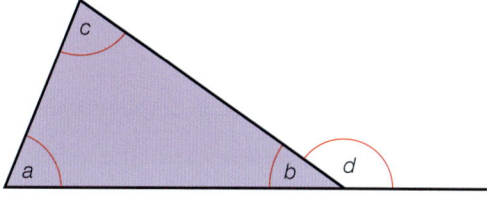

We already know that:

$a + b + c = 180°$ (angles of a triangle add up to 180°)

and that:

$b + d = 180°$ (angles on a straight line add up to 180°)

As the right-hand side of each equation is 180°, the left-hand sides must be equal to each other too, that is:

$a + b + c = b + d$

Subtracting b from each side gives:

$a + c = d$

In other words, the exterior angle of a triangle is equal to the sum of the two interior opposite angles.

> **Worked example**
>
> Calculate the size of angle x in this diagram.
>
> $65° + x = 110°$ (exterior angle of a triangle is equal to the sum of the two interior opposite angles)
>
> Therefore $x = 110° - 65° = 45°$.
>
>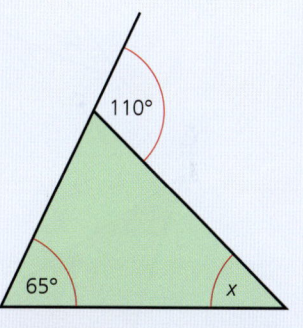

SECTION 2

Exercise 17.9

Calculate the size of each of the unknown angles in these diagrams.
Justify your answers using mathematical language.

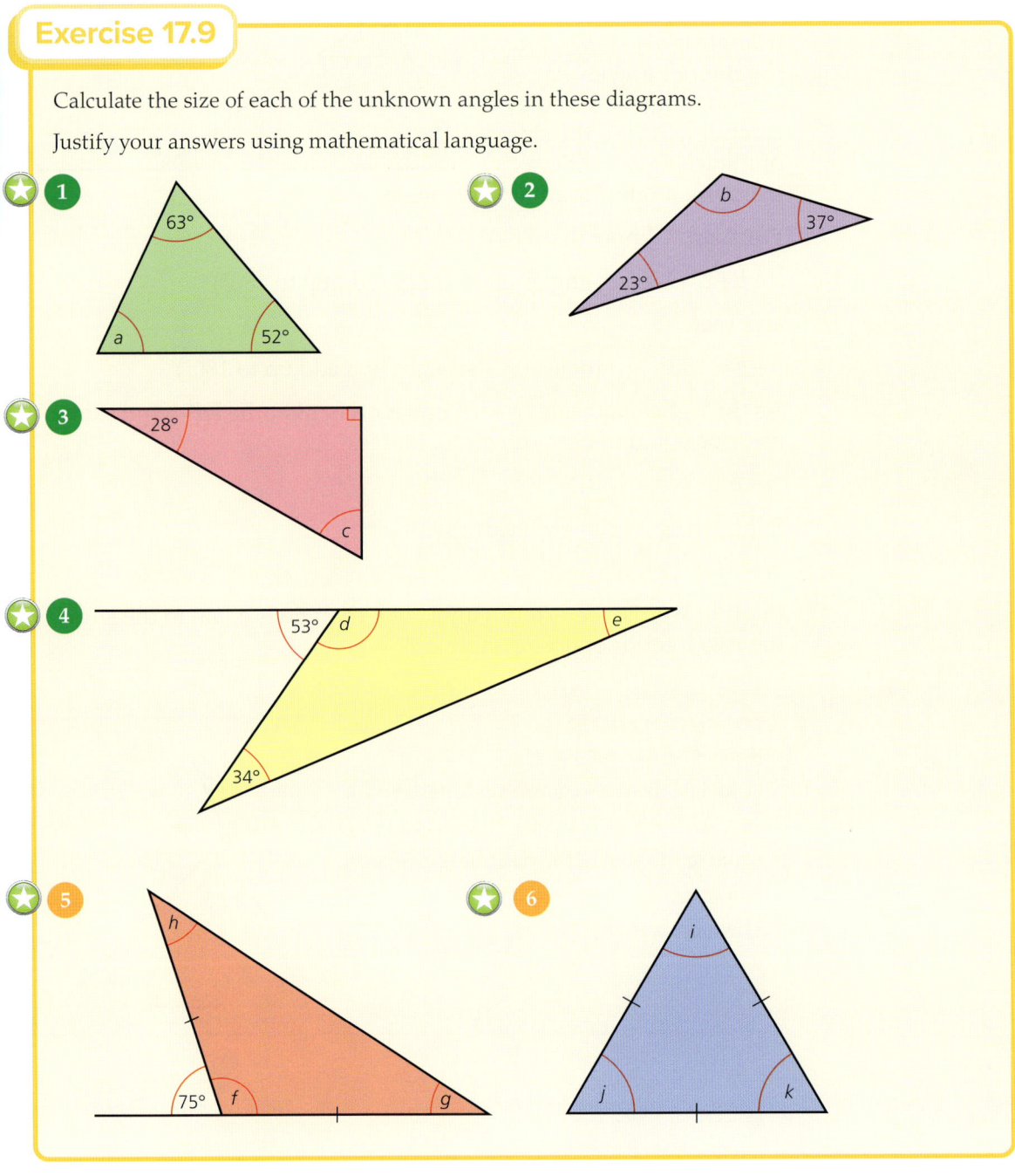

17 Constructions, lines and angles

Angles of a quadrilateral

A quadrilateral has four angles as shown.

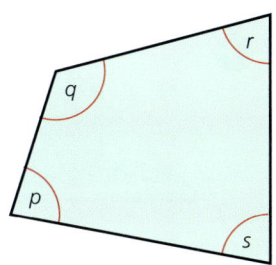

As with triangles, the sizes of the individual angles can vary, but the sum of the four angles is the same for any quadrilateral. The four angles always add up to 360°.

In this quadrilateral, for example,
$p + q + r + s = 360°$.

Exercise 17.10

1. That the four angles of any quadrilateral total 360° can be proved. The first part of the proof is given below. Copy and complete it. Any quadrilateral can be split into two triangles as shown:

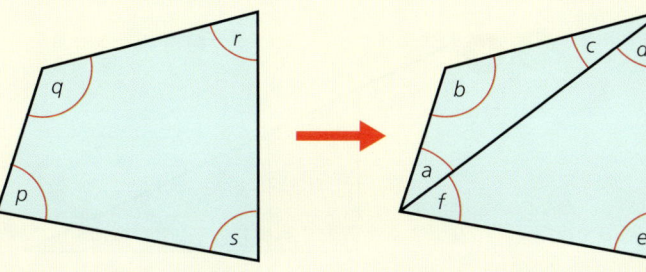

$a + b + c =$ .... because ....
$d + e + f =$ .... because ....
Therefore $a + b + c + d + e + f =$ ....
Complete the proof to show that $p + q + r + s = 360°$.
In questions 2–8, calculate the size of each of the unknown angles.
Justify your answers using mathematical language.

2.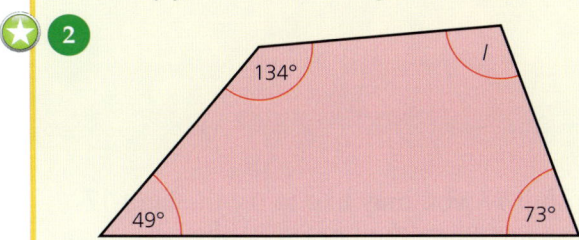

SECTION 2

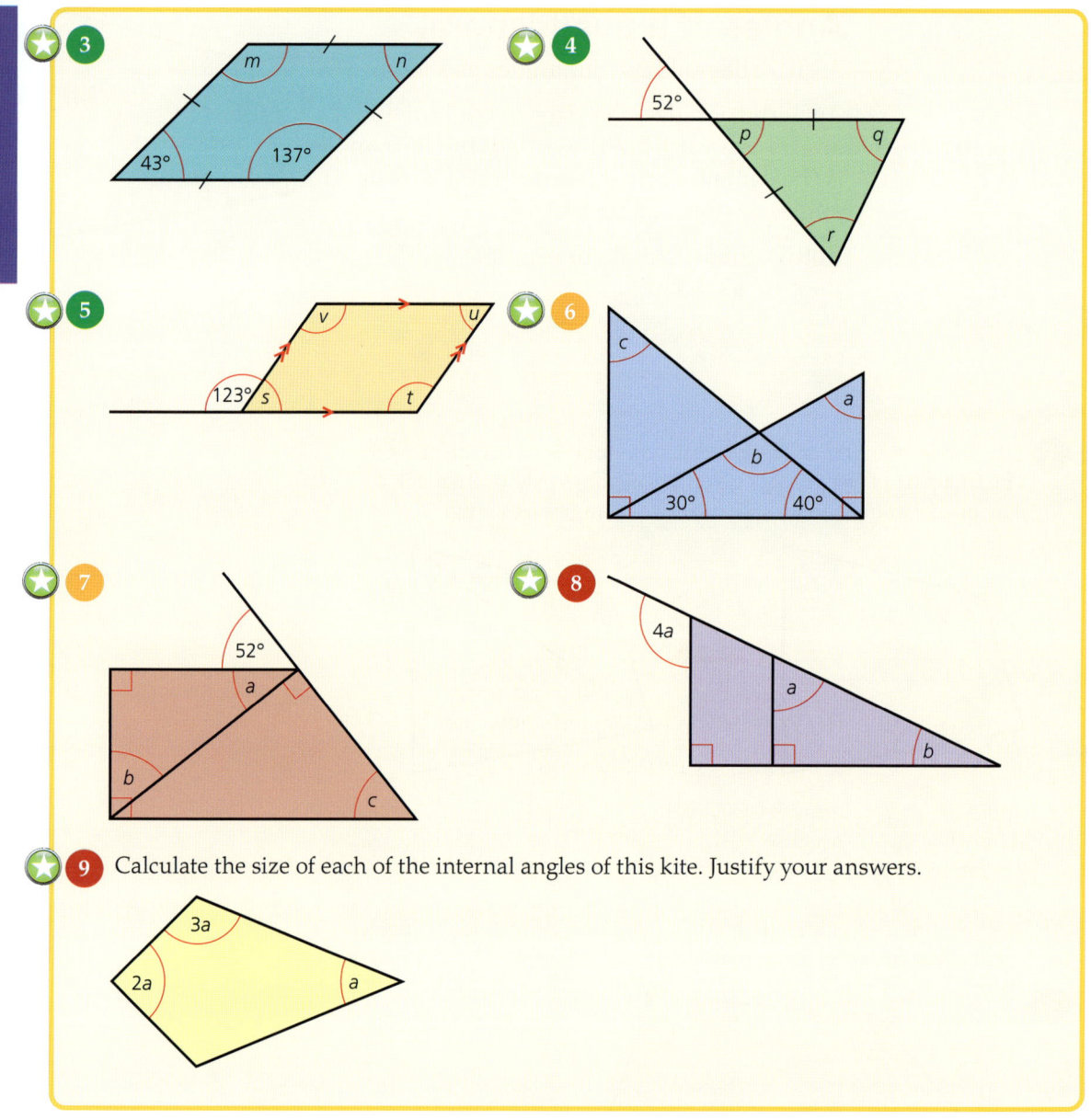

9 Calculate the size of each of the internal angles of this kite. Justify your answers.

Now you have completed Unit 17, you may like to try the Unit 17 online knowledge test if you are using the Boost eBook.

18 Algebraic expressions and formulae

- Understand that a situation can be represented either in words or as an algebraic expression, and move between the two representations (linear with integer or fractional coefficients).
- Understand that a situation can be represented either in words or as a formula (mixed operations), and manipulate using knowledge of inverse operations to change the subject of a formula.

Algebraic expressions

You will already be familiar with expressions and how to form them from given information.

For example, the rectangle below has dimensions as shown:

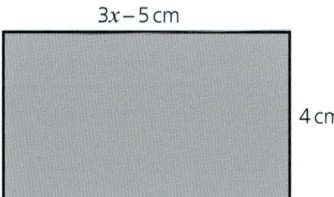

An expression for the perimeter is $(3x-5)+(3x-5)+4+4$

This can be simplified to $6x-2$ cm

An expression for the area is $4(3x-5) = 12x-20$ cm²

This unit builds on this by introducing more complex examples.

Exercise 18.1

1. The triangle opposite has the properties and dimensions shown.
   a Write an expression for the following
      i) the perimeter
      ii) the area.
   b If $m = 5$, calculate the perimeter and area of the triangle.

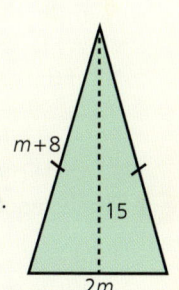

2. Two boxes, A and B, contain different coloured balls, either red (R) or yellow (Y).
   Box A contains 2R red balls and 4Y yellow balls.
   Box B contains 3R red balls.
   a Write an expression for the total number of balls in box A.
   b Write an expression for the total number of balls in boxes A and B combined.
   c If R = 4 and Y = 6, calculate the total number of balls in both boxes.

175

SECTION 2

3 Three friends, Gaby, Joe and Corri, all collect free gift tokens issued in a magazine. Gaby has x tokens, Joe has ten more than half the number Gaby has, while Corri has double Joe's amount. Four expressions are given below:

$\frac{1}{2}(x+10)$ $\frac{1}{2}x+10$ $x+20$ $2x+10$

   a Decide which expression represents the number of tokens owned by Joe and which represents the number of tokens owned by Corri.
   b Explain why Gaby must have an even number of tokens.
   c Write an expression for the total number of tokens they have.
   d If Joe has 30 tokens, how many do they have in total?

4 Two rectangles are shown below:

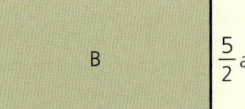

   a Write an expression for the perimeter of each rectangle.
   b Write an expression for the area of each rectangle.
   c Calculate the perimeter and area of each rectangle if $a = 6$ and $b = 2$.

5 A cuboid has dimensions as shown opposite.
Write an expression in simplified form for
   a its volume
   b its total surface area
   c the total length of its edges.
   d What value must x be greater than? Justify your answer.

6 An L-shaped object is shown. It is made from three squares of equal size.
Two congruent L-shaped pieces are arranged edge-to-edge.
   a Write an expression for the maximum possible perimeter of the combined shape.
   b Write an expression for the minimum possible perimeter of the combined shape.
   c **i)** Which of the two arrangements will have the greater area? Justify your answer.
      ii) Write an expression for this area.

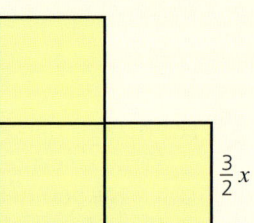

7 Three generations of a family (boy, mother and grandfather) have the following ages:

$m+2$ $100-m$ $3\frac{1}{2}m-15$

   a Write an expression for the total age of all three family members.
   b **i)** If $m = 20$, which expression represents the age of the grandfather?
      ii) If $m = 35$, which expression represents the age of the grandfather?
   c Decide on a realistic value for m. Justify your answer.

18 Algebraic expressions and formulae

> The 2 and the π are known as constants because their values do not change.

Deriving formulae

You already know that a formula describes the relationship between different variables.

For example, the circumference of a circle is given by the formula $C = 2\pi r$, where both C and r are variables. Their values can change and are related to each other. Written as $C = 2\pi r$, C is the subject of the formula. To find r, the formula needs to be rearranged to make r the subject.

Worked example

A compound shape is made from a rectangle and two semicircles as shown:

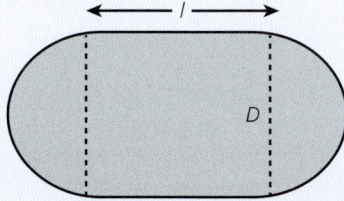

> A compound shape is a shape which can be broken down into simpler shapes.

i) Derive a formula for the perimeter, P, of the whole shape.

The perimeter of the two semicircles is equal to the circumference of one complete circle, i.e. πD.

The total length of both sides of the rectangle is $2l$.

Therefore, the perimeter $P = \pi D + 2l$.

ii) Rearrange the formula to make D the subject.

$P = \pi D + 2l$

$P - 2l = \pi D$ (subtract $2l$ from both sides)

$\frac{P - 2l}{\pi} = D$ (divide both sides by π)

> Rearranged, the formula can be written as $\frac{P-2l}{\pi} = D$ or as $D = \frac{P-2l}{\pi}$. In both cases D is the subject.

iii) If the perimeter of the shape is 50 cm and the side length of the rectangle 15 cm, calculate to 1 decimal place the diameter of the semicircle.

Using $D = \frac{P-2l}{\pi}$, the values for P and l can be substituted into the formula:

$D = \frac{50 - 2 \times 15}{\pi} = \frac{20}{\pi}$

Therefore, $D = 6.4$ cm (1 decimal place).

> Note that sometimes the answer can be left as $\frac{20}{\pi}$ which is known as leaving your answer in **exact form**.

> **LET'S TALK**
> Why is leaving an answer as $\frac{20}{\pi}$ more accurate than giving it as 6.4 (1 decimal place)?

SECTION 2

Exercise 18.2

1 Rearrange the following formulae to make the variable in **red** the subject.
- a $A = B - 3$
- b $P = 3R - 12$
- c $D = 2E + 4$
- d $L = \dfrac{M}{2}$
- e $T = \dfrac{V}{3}$
- f $F = \dfrac{1}{2}G + 1$

2 Rearrange the following formulae to make the variable in **red** the subject.
- a $R = 5C + 2$
- b $A = \dfrac{7b}{2}$
- c $P = \dfrac{Q - 4}{3}$
- d $X = \dfrac{1}{2}Y - 2$
- e $M = 2(P + 6)$
- f $D = \dfrac{4}{3}E$

3 A rectangle with dimensions $2a$ and $2b$ has another rectangle with half those dimensions joined to its right side as shown.

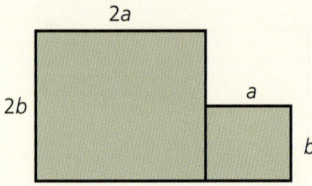

- a Show that the formula for the perimeter (P) of the whole shape is $P = 2(3a + 2b)$.
- b Another rectangle, also with dimensions half those of the original one, is attached to the left side as shown.

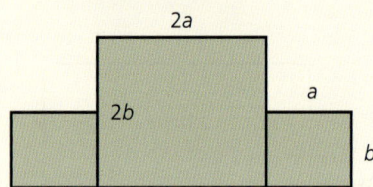

  Derive the formula for the perimeter of this shape.
  Give your answer in factorised form.
- c If $a = 10$ cm and $b = 6$ cm, use your formulae to calculate the perimeter of
    i) the first two rectangles
    ii) the three rectangles.
- d If the perimeter of the three rectangles is 160 cm, calculate a possible value for a and b.

4 A family is moving home and is looking at the different pricing options offered by a removal company. These are:
Option 1: $150 loading fee and then an additional $0.80 for each km of transport.
Option 2: $75 loading fee and then an additional $1.30 for each km of transport.
- a Write a formula for the cost (C) of moving the house contents a distance (d km) for both options.
- b Which option would you recommend the family choose? Justify your answer fully, showing any calculations clearly.

5 A trapezium is split into two triangles, X and Y, as shown.

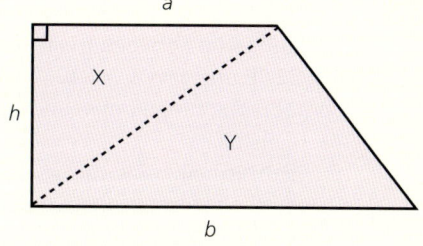

  a Write an expression for the area of triangle X.
  b Show that the area (A) of the trapezium is given by the formula $A = \frac{1}{2}(a+b)h$.
  c Calculate the area of the trapezium if $a = 12$ cm, $b = 18$ cm and $h = 6$ cm.
  d To calculate the value of b, the formula can be rearranged as $b = \frac{2A}{h} - a$.
     Calculate the value of b when $A = 27\,\text{cm}^2$, $h = 6$ cm and $a = 2$ cm.

6 The diagrams show polygons with different numbers of sides. One vertex of each polygon is connected to every other vertex of the polygon, dividing it into triangles as shown.

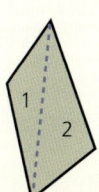 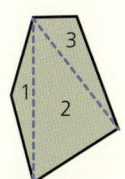 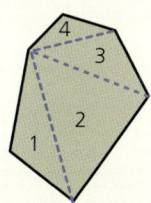 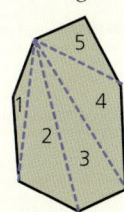

  a Copy the table below for the number of sides and the number of triangles in each polygon.

Number of sides	3	4	5	6	7
Number of triangles					

  b Write a formula linking the number of sides (n) of a polygon and the number of triangles (t) it can be split into.
  c If a polygon has 100 sides, how many triangles can it be split into?
  d What do the sum of the interior angles of a triangle add up to?
  e For the following questions, deduce what the sum of the interior angles of each polygon must add up to. Justify your answer.
      i)   pentagon
      ii) hexagon
      iii) heptagon.

  > A heptagon is a polygon with 7 sides.

  > **LET'S TALK**
  > Can you think of other cases where the prefix 'hept-' is used to represent the number 7?

  f A mathematics textbook states that the formula for finding the sum (s) of the interior angles of a polygon is given by the formula $s = 180(n-2)$, where n is the number of sides of the polygon. On the next page of the book it says the formula is $s = 180n - 2$.
      i) Decide which one is correct.
      ii) What is the mistake made in the incorrect one?

  > A dodecagon is a polygon with 12 sides.

  g Use the correct formula above to calculate the sum of the interior angles of a dodecagon.
  h Rearrange the formula to make n the subject.
  i Calculate the number of sides a polygon has if the sum of its interior angles is 2880°.

SECTION 2

 7 Two of the units for measuring temperature are Celsius and Fahrenheit.
The relationship between the freezing and boiling points of water in both temperature scales is:
0 °C = 32 °F and 100 °C = 212 °F.
- a What is the range between the freezing point and boiling point of water in
  - i) Celsius
  - ii) Fahrenheit.
- b Using the information given so far, and your answers in part (a), explain why the formula to convert a temperature in degrees Celsius to its equivalent temperature in degrees Fahrenheit is given by:
  $F = 1.8C + 32$
- c Rearrange the formula to make C the subject.
- d Work out the following temperature equivalents:

Degrees Fahrenheit	Degrees Celsius
100	
0	
	160
	−273

LET'S TALK
1. Research the significance of the temperature −273 °C.
2. There is another temperature scale called the Kelvin. Research what it is, why it is used and how it relates to Celsius.

 Now you have completed Unit 18, you may like to try the Unit 18 online knowledge test if you are using the Boost eBook.

19 Probability experiments

- Design and conduct chance experiments or simulations, using small and large numbers of trials. Compare the experimental probabilities with theoretical outcomes.

Experiments and simulations

You are already familiar with the idea that the topic of probability is the study of chance. As chance is unpredictable, what is supposed to happen in theory (**theoretical probability**) will not necessarily happen in reality.

Worked example

A dice is rolled 60 times and 600 times, and the number shown recorded each time. The results are presented in the table below.

Number	1	2	3	4	5	6
Frequency 60 rolls	2	6	5	8	5	34
Frequency 600 rolls	105	92	98	104	88	113

Comment on whether you think the dice is biased. Justify your answer.

For an unbiased six-sided dice, each number is equally likely and therefore the theoretical probability of each number is $\frac{1}{6}$.

With 60 rolls we would expect each number to occur approximately 10 times.
- These results would suggest that the dice might be biased towards the number 6.

With 600 rolls we would expect each number to occur approximately 100 times.
- These results would suggest that the dice is not biased as they are all fairly close to 100.

As 600 rolls gives more reliable data than 60 rolls, we can conclude that the dice is unlikely to be biased.

LET'S TALK
How far off from 100 would the results have to be for you to consider the dice to be biased?

SECTION 2

Exercise 19.1

1. A coin is flipped 100 times and the number of times it lands on either heads or tails is recorded. The results are given in the table opposite.

	Heads	Tails
Frequency	56	44

   Comment, justifying your answer, whether you think the coin is biased towards heads.

2. a Roll an ordinary dice six times and record the results. Is the dice biased? Justify your answer.
   b Conduct an experiment to test whether the dice is biased or not. You will need to include the following points in your experiment write-up:
      - hypothesis
      - method
      - results
      - analysis
      - conclusion.

   > For a proper experiment you will need to make clear what it is you are trying to find out (hypothesis), how you are going to find it out (method), what happened in the experiment (results), what you did with the results (analysis) and a summary of your findings (conclusion).

3. a Draw a possible net of a cube.
   b Out of card, construct a cube of side length 4 cm.
   c Conduct an experiment to be able to **convincingly** answer the following question, 'Is my dice fair'?

   > Remember to account for tabs in order to be able to stick it together.

   LET'S TALK
   What do the numbers on opposite faces of a dice add up to?

4. a Construct, out of card, a regular hexagon using a pair of compasses and ruler as follows:

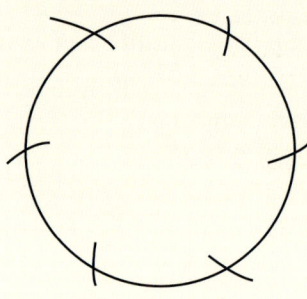 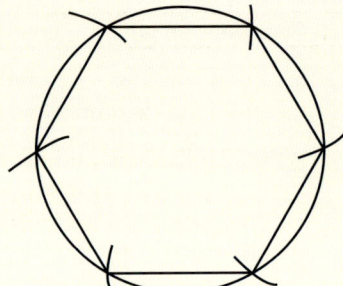

   - Open up a pair of compasses and draw a circle.
   - Keeping the compasses open by the same amount, put the point on the circumference of the circle and draw an arc. Ensure the arc intersects the circumference.
   - Place the compass point on the point of intersection of the arc and the circumference and draw another arc.
   - Repeat the above procedures until there are six arcs drawn (see the diagram above).
   - Draw lines joining each of the six arcs.
   - Cut out the hexagon and divide it into six equilateral triangles and number each one with the numbers 1–6.
   - Insert a short pencil through the centre to form your spinner.

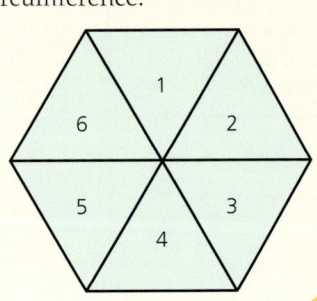

19 Probability experiments

b Conduct an experiment to be able to **convincingly** answer the question, 'Is my spinner fair'?

c i) If your spinner is biased, suggest **improvements** to try to correct the bias.
   ii) Test your **improvements**.

d i) If your spinner was fair, suggest modifications to make it biased towards a particular number.
   ii) Test your modifications.

5 You will need to work with an unbiased coin for this question and have access to an Excel spreadsheet.

a Flip the coin 100 times, each time recording 'heads' as a 1, and 'tails' as a 0 onto your spreadsheet. An example of how to set out your spreadsheet is shown below:

Spin no.	Heads (1) Tails (0)	Total no. of heads	Proportion of heads
1	1	1	1
2	0	1	0.5
3	0	1	0.333333333
4	0	1	0.25
5	1	2	0.4
6	1	3	0.5
7	0	3	0.428571429
8	1	4	0.5

To conduct this experiment efficiently you may need to research the use of formulae in Excel.

- Continue the numbers to 100 using a formula.
- Record your results in this column.
- Use a formula to keep a running total of 'heads'.
- Use a formula to calculate the proportion of 'heads' after each spin.

b Once you have completed your spreadsheet, plot a line graph to show how the experimental probability of heads changes as the number of spins increases.

c In your own words, explain what your graph shows.

 Now you have completed Unit 19, you may like to try the Unit 19 online knowledge test if you are using the Boost eBook.

183

20 Equations and inequalities

- Understand that a situation can be represented either in words or as an equation. Move between the two representations and solve the equation (integer or fractional coefficients, unknown on either or both sides).
- Understand that letters can represent open and closed intervals (two terms).

Further equations

An equation represents two quantities that are equal to each other. In Stage 7, you saw that, to help understand what an equation is and how it can be manipulated, it can be thought of as a pair of balanced scales.

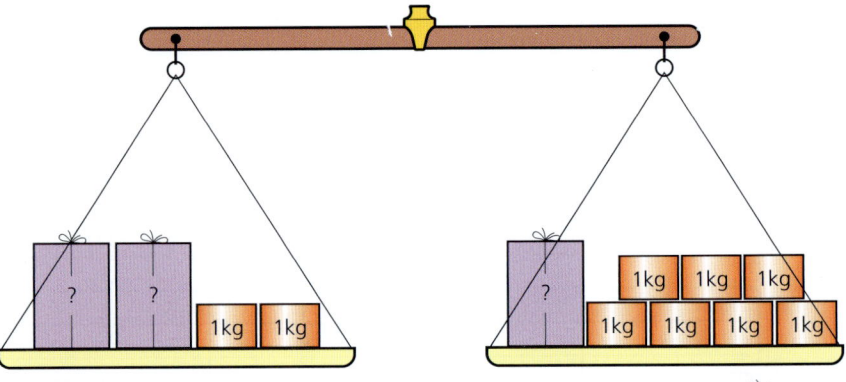

In the scales above there are two different types of object and .

The left-hand side of the scales balances the right-hand side, i.e. the total masses on both sides are equal.

20 Equations and inequalities

> **Worked example**
>
> For the scales on the previous page, find the mass of each ? if the mass of each 1kg is 1 kg.
>
> Subtracting 2 1kg from each side gives:

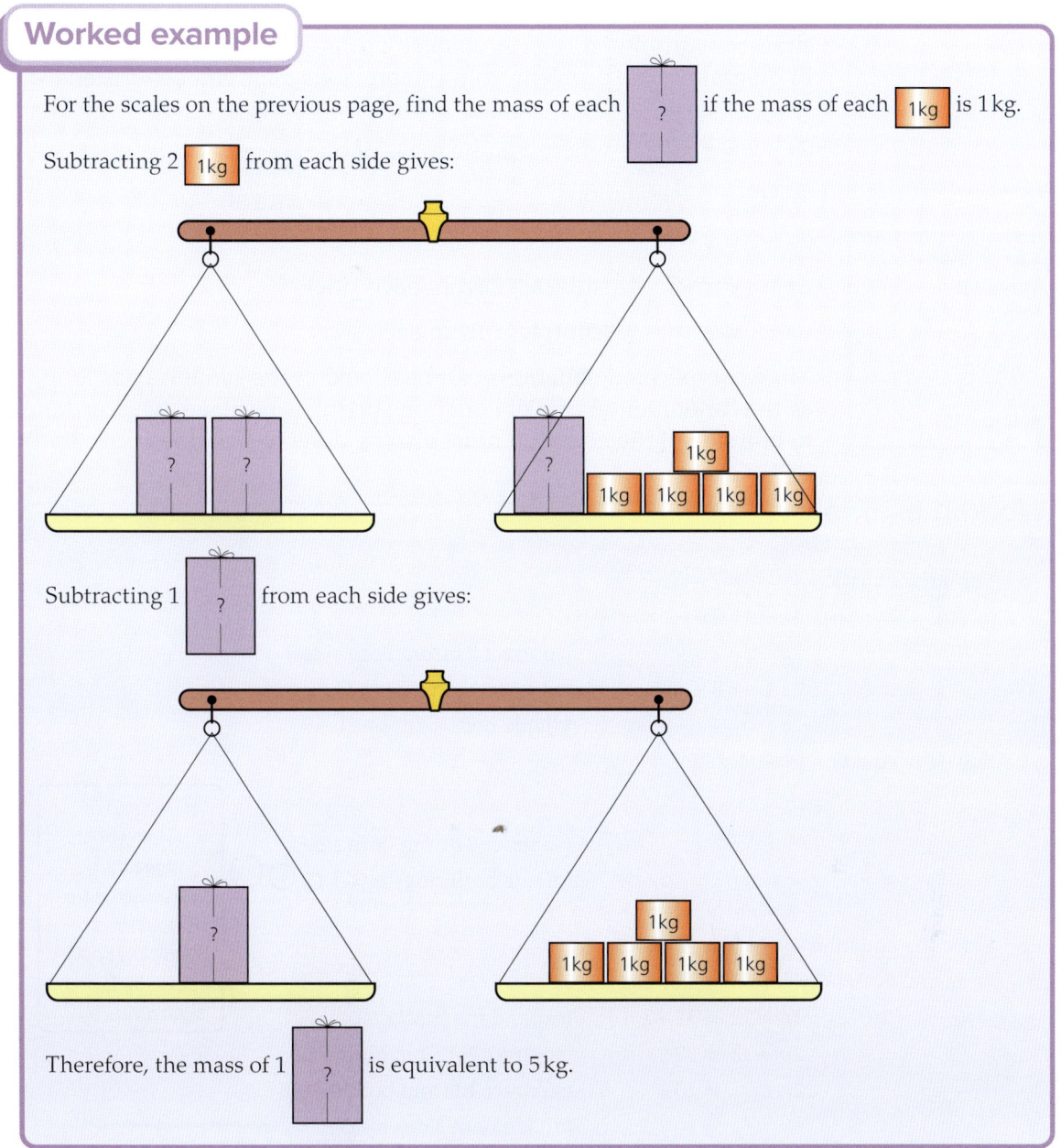

Subtracting 1 ? from each side gives:

Therefore, the mass of 1 ? is equivalent to 5 kg.

SECTION 2

However, drawing scales at each stage in the solution can be a time-consuming process. A quicker way is to use algebra instead of diagrams. The problem from the previous page can be written as:

$2x + 2 = x + 7$ where x is the mass of one 🎁.

To solve the equation:

$2x + 2 = x + 7$
$2x = x + 5$ (subtract 2 from each side)
$x = 5$ (subtract x from each side)

More complicated equations can be solved in the same way as long as the fundamental rule of equation solving is used, i.e. what is done to one side of the equation must also be done to the other.

LET'S TALK
Why is it important to do the same thing to both sides of an equation when solving it?

Worked example

a Solve the equation $4x - 6 = 2x + 2$.

$4x - 6 = 2x + 2$
$2x - 6 = 2$ (subtract $2x$ from both sides)
$2x = 8$ (add 6 to both sides)
$x = 4$ (divide both sides by 2)

b Solve the equation $15 = 24 + \frac{3}{2}x$.

$15 = 24 + \frac{3}{2}x$
$-9 = \frac{3}{2}x$ (subtract 24 from both sides)
$-6 = x$ (divide both sides by $\frac{3}{2}$)
$x = -6$

Instead of dividing both sides by $\frac{3}{2}$, an alternative would be to multiply both sides by 2 and then divide by 3.

c Solve the equation $4(m + 3) = 32 + 2m$.

$4(m + 3) = 32 + 2m$
$4m + 12 = 32 + 2m$ (expand the brackets)
$2m + 12 = 32$ (subtract $2m$ from both sides)
$2m = 20$ (subtract 12 from both sides)
$m = 10$ (divide both sides by 2)

d Solve the equation $3(x - 7) = 2(2x - 15)$.

$3(x - 7) = 2(2x - 15)$
$3x - 21 = 4x - 30$ (expand the brackets)
$-21 = x - 30$ (subtract $3x$ from both sides)
$9 = x$ (add 30 to both sides)
$x = 9$

LET'S TALK
How can you check that your answers are correct?

20 Equations and inequalities

Exercise 20.1

Solve these equations.

1.
   a. $4c = 3c + 4$
   b. $6c = 5c + 8$
   c. $8c = 7c + 3$
   d. $5c = 4 + 4c$
   e. $7c = 8 + 6c$

2.
   a. $3d = 2d - 2$
   b. $5d = 4d - 4$
   c. $6d = 5d - 9$
   d. $8d = 7d - 11$
   e. $10d = 9d - 9$

3.
   a. $3e = e + 4$
   b. $6e = 3e + 12$
   c. $5e = e + 8$
   d. $7e = 2e + 20$
   e. $11e = 4e + 21$

4.
   a. $4f = 2f - 6$
   b. $5f = 2f - 9$
   c. $8f = 4f - 24$
   d. $9f = 4f - 20$
   e. $12f = 7f - 35$

5.
   a. $5 = \frac{g}{2} + 9$
   b. $9 = \frac{g}{3} - 3$
   c. $8 = \frac{2}{3}g + 2$
   d. $14 = \frac{3}{2}g + 2$
   e. $28 = \frac{3}{4}g - 8$

6.
   a. $2(h+1) = 6$
   b. $3(h+2) = 18$
   c. $4(h+5) = 40$
   d. $7(h-3) = 14$
   e. $9(h-7) = 36$

7.
   a. $3 + \frac{3}{2}k = 4k - 2$
   b. $5k + 7 = \frac{3}{8}k + 81$
   c. $6k - 3 = \frac{3}{5}k + 24$
   d. $\frac{2}{3}k + 4 = 10k - \frac{2}{3}$
   e. $8 - 3k = \frac{k}{2} - 6$

8.
   a. $2(l-3) = 4(l-6)$
   b. $2(l+1) = 3(l-5)$
   c. $5(l-4) = 3\left(\frac{l}{2}+2\right)$
   d. $6(2l+3) = 5\left(\frac{4}{5}l+2\right)$
   e. $7\left(7+\frac{2l}{7}\right) = 3\left(9l-\frac{1}{3}\right)$

Constructing and solving equations

Being able to solve equations is one thing; in the real world, however, often it is necessary to construct the equation from the information first before going on to solve it.

Worked example

Amelia and Zach are playing a 'think of a number' game. Amelia says, 'I think of a number, subtract 5 and then double it. The answer is 28.' What is the number that Amelia chose?

Using n to represent the number Amelia chose, we can write:

$2(n-5) = 28$

Then we solve the equation as before:

$n - 5 = 14$

$n = 19$

So the number she chose is 19.

> The brackets could have been multiplied out first. In this case, though, dividing both sides by 2 was the chosen method. The final answer would be the same whichever method is used.

SECTION 2

Exercise 20.2

Construct an equation from the information given in each question 1–10 and then solve it.

1. A number is subtracted from 10 and the answer is 3. What is the number?
2. A number is subtracted from 3. The result is multiplied by 4 and the answer is 8. What is the number?
3. A number is subtracted from 7. The result is multiplied by 3 and the answer is 6. What is the number?
4. A number is subtracted from 9. The result is multiplied by 10 and the answer is 40. What is the number?
5. A number is subtracted from 16. The result is multiplied by 33 and the answer is 66. What is the number?
6. 6 is subtracted from a number and the result is multiplied by 4.
   The answer is the same if 3 is subtracted from the number and the result is multiplied by 2. What is the number?
7. If 5 is subtracted from a number and the result is multiplied by 3, the answer is the same as adding 1 to the number and doubling the result.
   What is the number?
8. Add 2 to a number and then multiply the result by 6. If this is the same as subtracting 4 from the number and then multiplying the result by 10, what is the number?
9. Multiply a number by 4 and add 2, then multiply the result by 10.
   The answer is the same if you multiply the number by 6, add 10 and then multiply the result by 5. What is the number?
10. 9 times a number has 1 subtracted from it. The result is trebled.
    The answer is the same if you add 7 to twice the number and multiply the result by 7. What is the number?

Exercise 20.3

1. Two rectangles A and B are shown in the figure below.
   The area of Rectangle B is twice the area of rectangle A.

   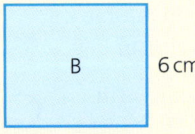

   2 cm   A
          $(x + 3)$ cm

          B   6 cm
          $(2x - 1)$ cm

   a. Write a formula for the area of A in terms of x.
   b. Write a formula for the area of B in terms of x.
   c. i) Write an equation from the information given.
      ii) Solve the equation to find the value of x.

20 Equations and inequalities

2 A composite shape can be split into two rectangles, P and Q, as shown:

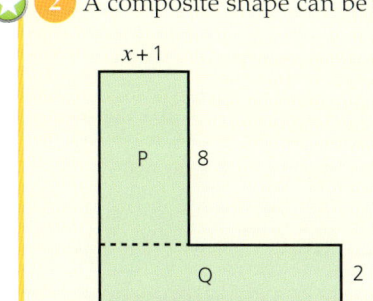

If the areas of P and Q are equal, calculate the area and perimeter of the composite shape.

3 A tile has a length of $3x + 2$ units and width of $2x - 1$ units.
Two of the tiles are placed side by side as shown:

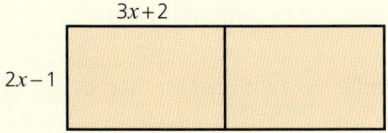

If the perimeter of both tiles is 86 cm, calculate the area of one tile.

4 Three cards, X, Y and Z, each have a number written on the back.

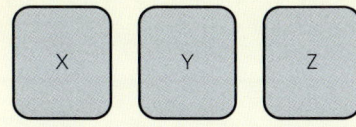

The number on X is ten times bigger than that on card Y.
The number on Z is $2\frac{1}{2}$ times bigger than that on X.
If the total of all three cards is 90, work out the number on the back of each card.

Inequalities

You are now familiar with the equals sign, used in equations to show that one thing is equal to another. There are a number of other signs which show that one thing is not equal to another. These are called inequality signs and they are used in inequalities.

The basic inequality signs are < and >.

$x < y$ means x is less than y.

$x > y$ means x is greater than y.

SECTION 2

So, a sentence such as 'Everyone in a class of students is less than 13 years old' could be expressed as:

$a < 13$, where a is the age of a student, in years.

Two more signs, $\geqslant$ and $\leqslant$, are formed by putting part of an equals sign under the inequality sign.

$x \leqslant y$ means x is less than or equal to y.

$x \geqslant y$ means x is greater than or equal to y.

So, a sentence such as 'This nursery school takes children who are at least four years old' could be expressed as:

$a \geqslant 4$ where a is the age of a child, in years.

The following pairs of inequalities mean the same:

$x < y$ (x is less than y)	and $y > x$ (y is greater than x)
$x \leqslant y$ (x is less than or equal to y)	and $y \geqslant x$ (y is greater than or equal to x)

So, $5 < x$ is usually written as $x > 5$.

Another sign used is $\neq$, which means is not equal to. So $x \neq 5$ means x is not equal to 5.

> Language is important when deciding what sort of inequality to use. 'At least' can also be written as 'greater than or equal to'.

Worked example

a Write the inequality $a \leqslant 8$ in words.
   $a \leqslant 8$ means a is less than or equal to 8, i.e. a is at most 8.
b Write the inequality $b \geqslant 5$ in words.
   $b \geqslant 5$ means b is greater than or equal to 5, i.e. b is at least 5.

Exercise 20.4

Write these inequalities in words.

1 a $a < 6$ b $b > 5$ c $c \neq 10$
2 a $a \leqslant 7$ b $b \geqslant 3$ c $c \leqslant 10$
3 a $4 < a$ b $7 > b$ c $c \neq 8$
4 a $8 > a$ b $5 < b$ c $5 \neq c$
5 a $6 \geqslant a$ b $9 \leqslant b$ c $3 \neq c$

> **LET'S TALK**
> What other ways can language be used to describe these inequalities? Is your answer the only way of describing the same thing?

20 Equations and inequalities

Combined inequalities

It is possible to combine two inequalities. For example, $p > 7$ and $p < 10$ can be combined as $7 < p < 10$, which means that p is greater than 7 but less than 10.

$7 \leqslant q \leqslant 10$ means that q is greater than or equal to 7 but less than or equal to 10.

Exercise 20.5

Write down the integer values represented by each of the following inequalities.

1. $3 < a < 6$
2. $6 \geqslant b \geqslant 1$
3. $7 \leqslant c \leqslant 12$
4. $5 > d > 0$
5. $3 < e \leqslant 8$
6. $8 \geqslant f > 5$
7. $-4 < g < -2$
8. $-7 \geqslant h > -10$
9. $-2 \leqslant i < 3$
10. $4 > j \geqslant -1$

11. The numbers $-2, -1, 0, 1, 2, 3, 4, 5$ represent all the integer values represented by an inequality.
    a Write down a possible inequality which represents these values.
    b Are there other inequalities which could represent the same integer values?
      If so, write down a possible alternative.

12. Olafur and Laure go to the same youth club.
    Olafur lives 8 km from the youth club.
    Laure lives 6 km from the youth club.
    a If the distance between Olafur and Laure's house is d km, write as an inequality the minimum and maximum distances that d could be.
    b Justify your answer.

LET'S TALK
How many other inequalities can you think of which represent the same integer values?

Now you have completed Unit 20, you may like to try the Unit 20 online knowledge test if you are using the Boost eBook.

Section 2 – Review

1. A theatre has exactly 460 seats.
   The ticket desk says that they have sold 400 tickets to 1 significant figure.
   Is there a possibility that they will not have enough seats for the number of tickets sold? Justify your answer.

2. Two classes take the same mathematics test. One class has students which have been set by ability, the other is a mixed ability class. The table below shows a summary of the test scores for each class.

	Mean	Median	Mode	Range
Class X	75	72	70	15
Class Y	68	70	94	82

   a Class X say that on average they did better than class Y. Comment on the accuracy of this statement.
   b Which class is likely to be the one set by ability? Justify your choice.

3. Give the equation(s) of the line(s) of symmetry in the diagram.

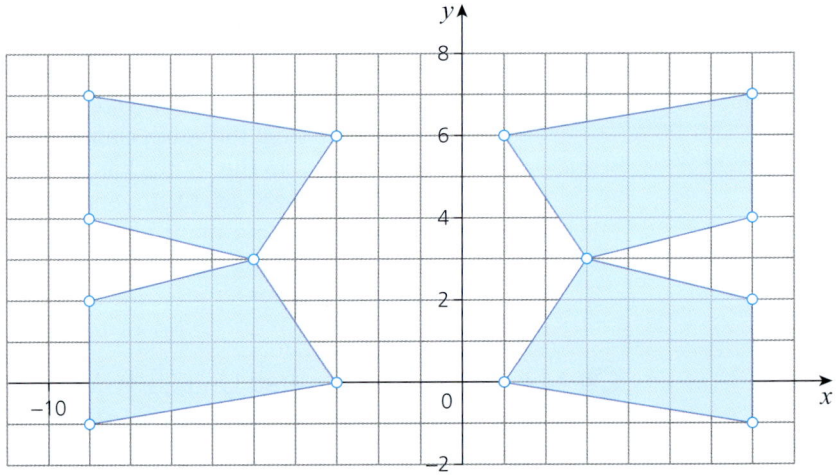

Section 2 – Review

4 Below are four mixed numbers and four improper fractions:

$$3\tfrac{1}{5} \quad 5\tfrac{1}{5} \quad 2\tfrac{2}{5} \quad 5\tfrac{4}{5} \quad \tfrac{8}{5} \quad \tfrac{16}{5} \quad \tfrac{29}{5} \quad \tfrac{26}{5}$$

  a Three of the mixed numbers can be paired with three of the improper fractions.
     Which ones form pairs?
  b Write the remaining mixed number as an improper fraction.
  c Write the remaining improper fraction as a mixed number.

5 A large rectangle has a smaller one cut from it as shown:

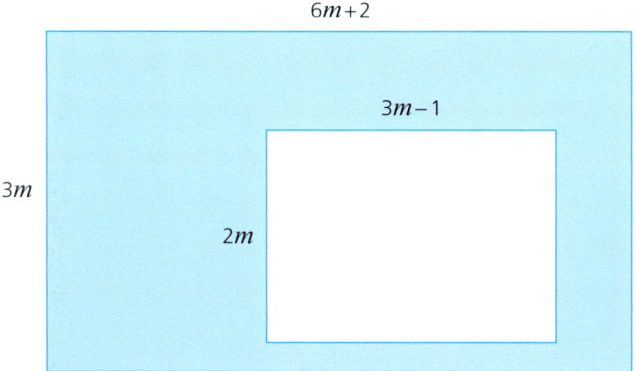

   Calculate the shaded area, giving your answer in simplified form.

6 Three ordinary coins are flipped.
  a Draw a tree diagram to show all the possible outcomes.
  b Highlight on your tree diagram all the outcomes where two or more of the coins land on heads.
  c What is the probability of two or more of the three coins landing on heads?

7 Using a ruler and a pair of compasses and leaving the construction lines visible:
  a Construct a triangle ABC where AB=5 cm, AC=3 cm and BC=7 cm.
  b Construct a line which bisects angle A of the triangle. Label the intersection with side BC, point X.

8 The masses in kg of three koalas are given in ascending order below.

$$\tfrac{7}{2}m - 1 \quad 2m+4 \quad m+8$$

  a Write an expression for the combined mass of all three koalas.
  b If the difference in masses between the lightest and middle koala is the same as the difference in mass between the middle and heaviest koala, calculate
     i) the value of m
     ii) the mass of each koala.

SECTION 2

9 A student decides to test if a coin is biased. He flips the coin four times and gets heads all four times.
   a Comment on whether you think the coin is biased.
   b Would you conduct the experiment differently? If so, how?
10 A number has 4 added to it and the result doubled. This produces the same answer as when the number is multiplied by 4 and then has 2 subtracted from it.
   What is the initial number?

SECTION 3

History of mathematics – The Persians (from 700CE)

'Mathematics is the language of the universe.'

Galileo Galilei

In Stage 7, you heard of Abu Ja'far Muhammad ibn Musa al–Khwarizmi. He is called 'The father of algebra' and was born in Baghdad in 790CE. He wrote the book *Hisab al–jabr w'al–muqabala* from which we have the word 'algebra'. He also introduced the decimal system from India. In 830CE Baghdad had the greatest university in the world and the greatest mathematicians studied there.

Algebra is simply a way of finding the general rules of arithmetic.

Muhammad Al Kharki (also known as Muhammad Al-Karaji) lived in the 11th century and did work on the theory of indices. He also worked on an algebraic method of calculating square and cube roots. He may also have travelled to the University in Granada, Spain where works of his are kept in the university library.

The poet Omar Khayyam is known for his long poem *The Rubaiyat*. He was also a fine mathematician. He also introduced the symbol 'shay', which became 'x'.

The beautiful mosques and palaces (see the image above) built by the Arabs and the Turks could only have been planned and built using mathematics. The intricate tiles found within these buildings have complex geometric designs.

21 Describing sequences

- Understand term-to-term rules, and generate sequences from numerical and spatial patterns (including fractions).
- Understand and describe nth term rules algebraically (in the form $n \pm a$, $a \times n$ or $an \pm b$, where a and b are positive or negative integers or fractions).

Term-to-term rules

A sequence is an ordered set of numbers. Each number in the **sequence** is called a term.

The terms of a sequence form a pattern. Below are examples of three different types of sequences.

- 2   4   6   8   10   12

In this sequence we are adding 2 to each term in order to produce the next term.

To find the 7th term in the sequence (the term in position 7), simply add 2 to the 6th term (the term in position 6).

- 1   3   9   27   81   243

In this sequence we treble each term in order to produce the next term.

- 1   3   6   10   15   21

Here, the difference between consecutive terms increases by 1 each time. It is also the sequence of **triangle numbers**.

There are two main ways of describing sequences. **Term-to-term rules** explain how to get from one term to the next. The rule for the **nth term** links the position number to the term itself.

Sometimes the patterns in sequences can be more difficult to spot and describe. This unit will build on what you covered in Stage 7 and look at slightly harder examples.

> **LET'S TALK**
> What other well-known sequences of numbers can you think of? Have they got special names?

21 Describing sequences

Worked example

a Consider the sequence below:

$\frac{13}{2}$ 5 $\frac{7}{2}$ 2 $\frac{1}{2}$

i) Describe the term-to-term rule.

The term-to-term rule is $-\frac{3}{2}$.

ii) Calculate the 7th term.

Position	1	2	3	4	5	6	7
Term	$\frac{13}{2}$	5	$\frac{7}{2}$	2	$\frac{1}{2}$	−1	$-\frac{5}{2}$

$-\frac{3}{2}$ $-\frac{3}{2}$

To calculate the 7th term we can continue the term-to-term rule.

> **LET'S TALK**
> What might be a drawback of using term-to-term rules for generating a specific term of a sequence?

b The stick diagrams below form a pattern.

Position   1   2   3

Pattern

Number of sticks 3   5   7

i) Draw the next two diagrams in the sequence.

ii) Describe the term-to-term sequence of the number of sticks in words and justify it by relating it to the pattern.

The term-to-term rule is +2, because each time just two sticks are added to complete the next triangle.

iii) Without drawing the pattern, calculate the number of sticks needed to complete the diagram with 8 triangles.

Using just the numbers and the term-to-term rule we can work it out as follows:

Position	1	2	3	4	5	6	7	8
Number of sticks	3	5	7	9	11	13	15	17

+2 +2 +2

The diagram with 8 triangles would have 17 sticks.

> **LET'S TALK**
> If the question had asked for the number of sticks for the pattern with 100 triangles, is there a more efficient way of using the term-to-term rule without writing it out fully?

SECTION 3

Exercise 21.1

1 In each of the following, the first term of a sequence has been given and the term-to-term rule. Generate the next three terms of the sequence.
   a 1st term = 4   Term-to-term rule is multiply by $\frac{1}{2}$
   b 1st term = $\frac{1}{5}$ Term-to-term rule is add $\frac{3}{5}$
   c 1st term = 2   Term-to-term rule is multiply by 3 and subtract 2
   d 1st term = 32   Term-to-term rule is multiply by $-\frac{1}{4}$
   e 1st term = 16   Term-to-term rule is multiply by $\frac{1}{2}$ and add 2
   f 1st term = 5   Term-to-term rule is multiply by -1

2 In each of the sequences below, the terms increase or decrease by the same amount from one term to the next. Two of the terms are known.
   i) Work out the term-to-term rule.
   ii) Calculate the missing three terms.
   a ....., 5, ..., ..., 11
   b 8, ..., ..., ..., 14
   c ....., ..., 3, ..., -11
   d ....., 9.6, ..., ..., -3

3 A sequence of patterns uses white and red squares. The first three patterns are shown below.

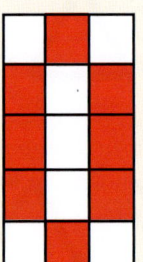  

Pattern 1   Pattern 2   Pattern 3

   a Which rule below gives the term-to-term rule for the number of white squares?
      i) multiply by 2 and subtract 4
      ii) add 3
      iii) subtract 2 and multiply by 2
   b Justify your answer to part (a).
   c Which rule below gives the term-to-term rule for the number of red squares?
      i) add 2
      ii) subtract 3 and multiply by 2
      iii) multiply by 2 and subtract 6
   d Justify your answer to part (c).

21 Describing sequences

LET'S TALK
Rather than writing out all 10 terms how can you work out the 10th term efficiently using the 1st term, the term-to-term rule and indices?

4 For each of the following sequences
  i) describe in words the term-to-term rule
  ii) write down the next two terms in the sequence and the 10th term.

a 3   6   12   24
b $\frac{7}{3}$ $\frac{8}{3}$ 3 $\frac{10}{3}$
c −16   −8   −4   −2
d 5   9   17   33
e $\frac{8}{5}$ $\frac{6}{5}$ $\frac{4}{5}$ $\frac{2}{5}$
f −1   2   −4   8

5 The stick diagrams below form a pattern sequence.

Pattern 1   Pattern 2   Pattern 3

a Draw the next diagram in the sequence.
b Describe the term-to-term rule in the sequence for the number of sticks and justify it by relating it to the patterns.
c Without drawing the pattern, calculate the number of sticks needed to complete a diagram with 20 squares.
d A pattern in the sequence is made from 175 sticks. How many squares are there?
e Explain why there will not be a pattern in the sequence made from 386 sticks.

6 A couple decide to save money in order to be able to buy a house. In the first month they save $5. In the second month they save three times that amount plus one extra dollar. In fact, they find that each month they manage to save three times the amount saved the previous month plus one extra dollar.
a How much do they save in the fourth month?
b How much have they saved in total in the fourth month?
c They need $18 000 for a deposit on a house. After how many months' saving can they put down the deposit?

SECTION 3

The nth term

Although calculating the term-to-term rule is a simple way of describing a sequence, it has its limitations if, say, the 100th term of a sequence is needed and only the first five terms are given.

Another way of describing a sequence is to write a rule that links a term to its position in the sequence. This is known as the rule for the **nth term**.

> **Worked example**
>
> a Consider the sequence below:
>   2   5   8   11   14
>   i) Describe the rule for the nth term of the sequence. To work out the rule for the nth term it is often easier to see the terms alongside their position in the sequence:
>
> | Position | 1 | 2 | 3 | 4 | 5 | n |
> |----------|---|---|---|----|----|-----|
> | Term | 2 | 5 | 8 | 11 | 14 | |
>
>   If each position number is first multiplied by 3 and then 1 is subtracted from the answer, we get the term. That is, $3 \times$ position number $- 1 =$ term. Written using n for the position gives the formula for the nth term; $3n - 1$.
>
>   ii) Use the rule for the nth term to calculate the 50th term.
>
>   For the 50th term the value $n = 50$ is substituted into the formula for the 50th term $= 3 \times 50 - 1 = 149$.
>
>   iii) The number 374 occurs in which position of the sequence?
>
>   This can be calculated by rearranging the formula for the nth term
>
> $374 = 3n - 1$
> $375 = 3n$ (add 1 to both sides)
> $n = 125$ (divide both sides by 3)
>
>   The 125th term is 374.
>
> b Look at the sequence of tiled patterns below.
>
>   Pattern
>   1   2   3
>
>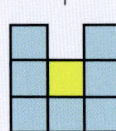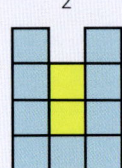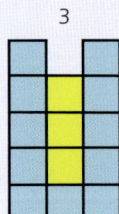

i) Write the rule for the nth term of the sequence for the number of blue tiles. Writing the results in a table gives:

Position	1	2	3	n
Term (blue tiles)	7	9	11	

The rule for the nth term of the sequence of blue tiles is $2n + 5$.

ii) Justify the rule for the nth term with reference to the tile patterns.

The position number is the same as the number of yellow tiles in each case. In each diagram doubling the number of yellow tiles can be represented by the dark blue tiles either side as shown.

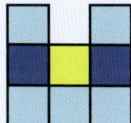 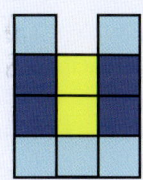 

LET'S TALK

If n is the number of yellow tiles, can you see how the total number of blue tiles can be worked out using this other method?

$n + 1$

$2(n + 1)$

$2(n + 1) + 3$

$2n + 5$

In each case there are five additional blue tiles.

Therefore, the total number of blue tiles is $2n + 5$.

Exercise 21.2

1 In each of the following sequences:
   i) Calculate the rule for the nth term.
   ii) Use your rule to calculate the 50th term.

a
Position	1	2	3	4	5	n
Term	$\frac{1}{2}$	1	$\frac{3}{2}$	2	$\frac{5}{2}$	

b
Position	1	2	3	4	5	n
Term	−2	−4	−6	−8	−10	

c
Position	1	2	3	4	5	n
Term	−5	−4	−3	−2	−1	

d
Position	1	2	3	4	5	n
Term	3	5	7	9	11	

LET'S TALK

In each of these questions can you see a relationship between the rule for the nth term and the term-to-term rule?

SECTION 3

e

Position	1	2	3	4	5	n
Term	$-\frac{1}{2}$	0	$\frac{1}{2}$	1	$\frac{3}{2}$	

f

Position	1	2	3	4	5	n
Term	0	−1	−2	−3	−4	

2 Two students, Ruth and Ricardo, are discussing the following sequence:

Position	1	2	3	4	5	n
Term	2	4	6	8	10	

> The nth term is also called the **general** term. When you find the nth term you are using your **generalising** skills.

Ruth states that the rule for the nth term is simply to double the position number. Ricardo says that another rule is just to add the position number to itself. Who is correct? Justify your answer using algebra.

3 Only two terms of a sequence are shown below:

Position	1	2	3	4	5	n
Term	3	6				

LET'S TALK
Are two terms of a sequence enough to work out the rule for the nth term?

a Which of the following statements is true?
  i) The rule for the nth term is $n+2$.
  ii) The rule for the nth term is definitely n^2+2.
  iii) The definite rule for the nth term cannot be worked out yet.
b Justify your answers to part (a).

4 In the previous exercise, you saw the following sequence of patterns of white and red squares.

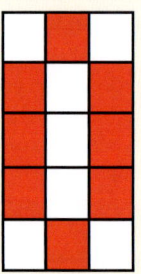  

Pattern 1   Pattern 2   Pattern 3

a Copy and complete the table of values.

Pattern number	1	2	3	4
Number of white squares				
Number of red squares				

b   i) Write a rule for the *n*th term for the number of white squares.
    ii) With reference to the patterns themselves, justify why your rule works.
c   i) Write a rule for the *n*th term for the number of red squares.
    ii) With reference to the patterns themselves, justify why your rule works.
d   Predict the number of white and red squares in the 85th pattern.

5   The following patterns of square tiles form a sequence as shown:

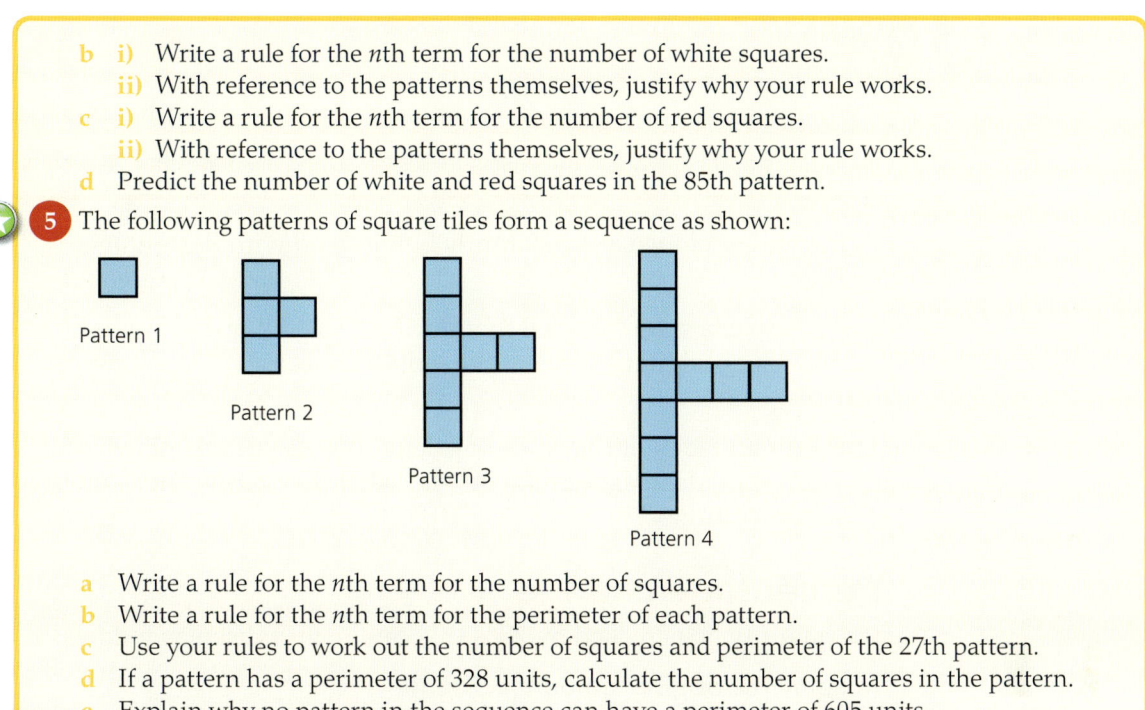

a   Write a rule for the *n*th term for the number of squares.
b   Write a rule for the *n*th term for the perimeter of each pattern.
c   Use your rules to work out the number of squares and perimeter of the 27th pattern.
d   If a pattern has a perimeter of 328 units, calculate the number of squares in the pattern.
e   Explain why no pattern in the sequence can have a perimeter of 605 units.

Now you have completed Unit 21, you may like to try the Unit 21 online knowledge test if you are using the Boost eBook.

22 Percentage increases and decreases

● Understand percentage increase and decrease, and absolute change.

In the real world we are constantly being faced with percentage change.

For example, a shop may say that during a sale all prices are reduced by 40%. A boss might say that the workforce is to receive a pay increase of 2%.

This unit will look at the ways of calculating these percentage increases and decreases and work out what they represent as an **absolute change**.

Percentage change

If a quantity is increased by 20% this can be visualised as follows:

Imagine the original amount being 100%.

A 20% increase therefore means that we have now 120% compared to the original amount.

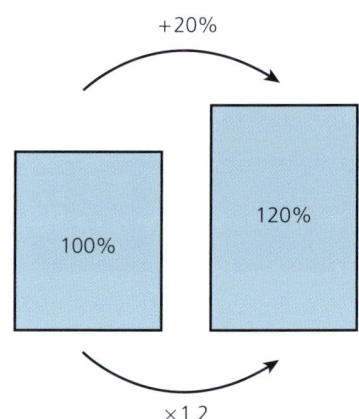

The diagram shows that an increase in 20% can be done in two ways:
1 By adding 20%
2 By multiplying by 1.2 because $100 \times 1.2 = 120$

The second method is particularly useful if we need to work out the actual amount after a percentage increase.

22 Percentage increases and decreases

Worked example

a A baby weighs 2500 g at birth. After 3 months her mass has increased by 60%.

i) Calculate her mass after 3 months.

The diagram to the right helps to visualise the problem.
The baby's original mass of 2500 g can be considered as 100%.
A 60% increase involves multiplying by 1.6 as $100 \times 1.6 = 160$.
As the original percentage is multiplied by 1.6 then the original mass must also be multiplied by 1.6. Therefore, the new mass $= 2500 \times 1.6 = 4000$ g (or 4 kg).
Note that to work out the multiplier, simply divide the new percentage by the original percentage, i.e. $\frac{160}{100} = 1.6$.

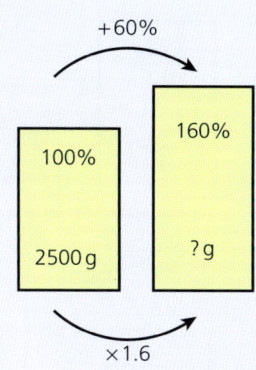

ii) What is the absolute change in the baby's mass?

This is the difference between the new mass and the original mass.
Therefore, the absolute change = 4000 g − 2500 g = 1500 g.

b A telephone helpline receives on average 120 calls a day.
One day, however, there is 500% increase in calls.
How many calls did they receive that day?
Using the diagram once again to visualise the problem.
The original number of calls can be represented by 100%.

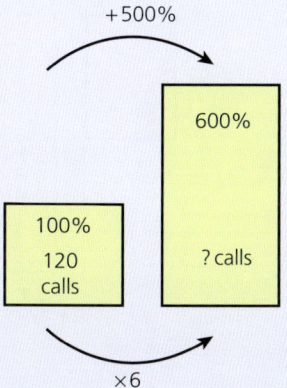

A 500% increase therefore leads to a value of 600%. The multiplier is calculated as $\frac{600}{100} = 6$.
The original number of calls is therefore also multiplied by 6.
So, the new number of calls is $120 \times 6 = 720$.
Note that an increase of 500% is the same as a multiplier of 6.

> **LET'S TALK**
> What is the relationship between the percentage increase and the multiplier?

SECTION 3

c A shop has a sale in which the prices of all clothes are reduced by 30%.

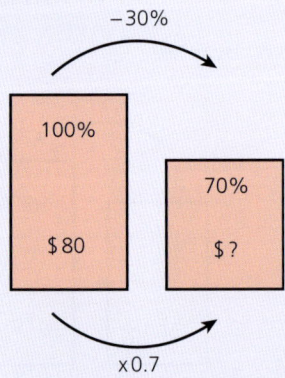

If a coat costs $80 before the sale, calculate its sale price.

To visualise the problem, see the diagram above.

The original price can be represented as 100%.

Therefore, a 30% decrease means that the sale price is 70% of the original price.

The multiplier is calculated as before by dividing the new percentage by the original one:

$\frac{70}{100} = 0.7$

Therefore, the sale price is $80 × 0.7 = $56.

A common mistake is to think a 30% reduction means multiplying by 0.3.

d Before going on a holiday, a couple have $600 in their bank account. During the holiday the amount in their account is reduced by 250%.

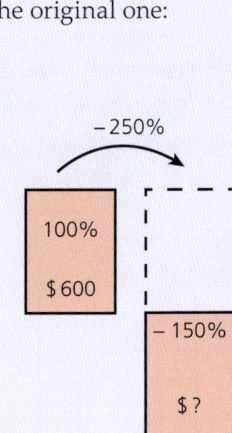

i) Calculate how much they owe the bank when they come back from their holiday.

To visualise the problem use the diagram to the right.

If the original amount is represented by 100% then a reduction of 250% leads to −150%.

The multiplier is the new percentage divided by the original percentage: $\frac{-150}{100} = -1.5$.

The amount in their bank account is therefore $600 × (−1.5) = −$900.

They therefore owe the bank $900.

ii) What is the absolute change in their account?

This is the difference between the new amount and the original amount.

Therefore, absolute change = −$900 − $600 = −$1500.

22 Percentage increases and decreases

Exercise 22.1

1. In each of the following calculate the new value when
   a. 200 is increased by 20%
   b. 350 is increased by 60%
   c. 820 is increased by 100%
   d. 1420 is increased by 2%
   e. 5050 is increased by 6%
   f. 27 is increased by 450%
   g. 50 is increased by 508%

   You may find it useful to use a diagram as in the two examples above to help answer these questions.

2. What multiplier corresponds to a 100% percentage increase?
   Justify your answer using an appropriate diagram.

3. The height of a sunflower is initially measured as 1.85 m. It continues to grow and at its full height it is found to have grown a further 18%.
   What is the absolute change in height? Give your answer in centimetres.

4. Two shops sell the same product. Shop A normally sells it for $120, while shop B normally sells it for $150.
   During a sale, shop A reduces the price by 30%, while shop B reduces it by 40%.
   Shop B claims that they are offering the biggest discount and so shoppers should buy the product from them. Is shop B correct? Give a **convincing** reason for your answer.

5. In a test a student gets 45 marks out of 80, which means that she has to resit.
   The pass mark for the retest is 80%.
   In the retest, the student improved her original mark by 40%.
   a. The student claims she has now passed the test. Is she correct? Show all of your working.
   b. How many marks did her score improve by?

6. What percentage change is represented by the following multipliers?
   State clearly whether it is an increase or decrease.
   a. ×1.5
   b. ×1.75
   c. ×(−2.1)
   d. ×0.82
   e. ×0.1
   f. ×8.4
   g. ×0.03
   h. ×(−0.3)

   You may find it useful to use a diagram to help visualise these multipliers.

7. In 2020 a factory worker receives a salary of $360 per week.
   In 2021 his pay is increased by 5%.
   If he works 40 hours per week, calculate the change in his hourly pay rate.

SECTION 3

⭐ 8 A rectangle is drawn with the following dimensions:

Height = 20 cm

Width = 80 cm

The height remains the same, but the width is increased by 700%.
What is the absolute change in area of the rectangle?

9 The temperature on a mountain at midday is 12 °C.
At night the temperature drops by 250%.
  a Calculate the night-time temperature.
  b Calculate the absolute temperature change.

10 A cliff has a height of 38 m above sea level.
The sea off the cliff has a depth d m.
The seabed represents a 220% drop in height from the top of the cliff.
Calculate the depth d of the sea.

⭐ 11 a A cuboid has dimensions as shown.
If each length is increased by 25%, calculate the absolute change in the volume of the cuboid.

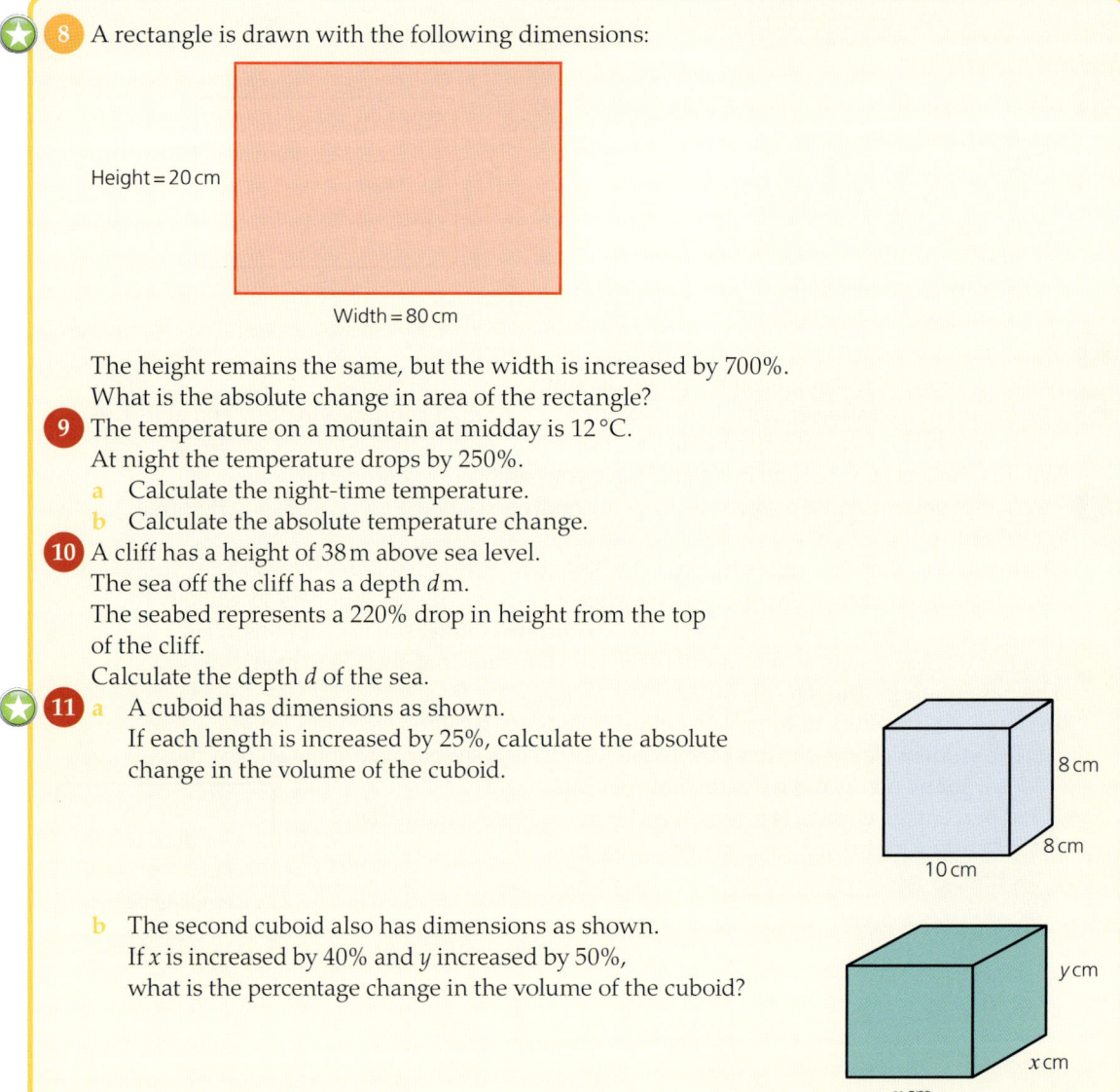

8 cm
8 cm
10 cm

  b The second cuboid also has dimensions as shown.
If x is increased by 40% and y increased by 50%,
what is the percentage change in the volume of the cuboid?

y cm
x cm
x cm

▶ Now you have completed Unit 22, you may like to try the Unit 22 online knowledge test if you are using the Boost eBook.

23 — 2D representations of 3D shapes

- Represent front, side and top view of 3D shapes to scale.

> An elevation is a two-dimensional side view of a three-dimensional object.

In Stage 7, you studied how to draw the different elevations of simple 3D objects.

These were either shapes made from cubes or simple 3D objects such as cylinders or triangular prisms.

This unit will look at the representation of 3D objects in 2D drawn to scale.

Elevations and scale

You will be familiar with how to draw 3D objects in 2D.

When you were younger, it is likely you will have produced a similar drawing to the one below.

> **LET'S TALK**
> What problems would a builder face if he tried to build this house from this child's drawing?

It is instantly recognisable as a house. It is in effect a front elevation (i.e. a view of the front) and the number of windows etc. can clearly be seen.

However, if this drawing was given to a builder and he was asked to build it, it would be quite difficult to do.

209

SECTION 3

In order to build a house from a drawing, the builder would possibly work from drawings similar to the ones below.

> **LET'S TALK**
> What scale is used on an architect's drawings? Are they always the same?

FRONT VIEW

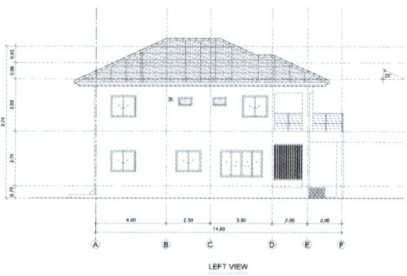
LEFT VIEW

BACK VIEW

RIGHT VIEW

These give the views from all four sides of the house. Importantly, they are also all in proportion to each other.

However, the diagrams above still do not have any dimensions. Full architectural drawings of a house will be drawn to scale and show all the dimensions. This unit will look at drawing elevations to scale.

> **Worked example**
>
> The shape below shows a simplified 3D view of a factory.
>
> It is made from a cuboid with dimensions 40 m × 20 m × 10 m and then two cubes, positioned as shown, both with a side length of 10 m.
>
> Using a scale of 1 : 1000 draw front and side and plan views of the factory.
>
> A scale of 1 : 1000 means 1 cm on the elevations represents 1000 cm (i.e. 10 m) in real life.

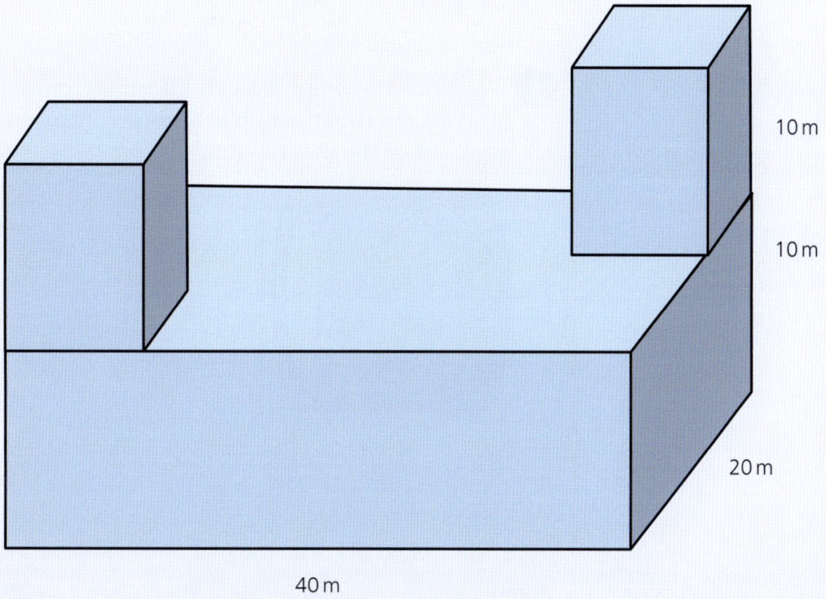

23 2D representations of 3D shapes

10 m

10 m

20 m

40 m

1 cm

Plan

2 cm

1 cm

Note how all the corresponding parts of the drawings align as shown by the dotted lines.

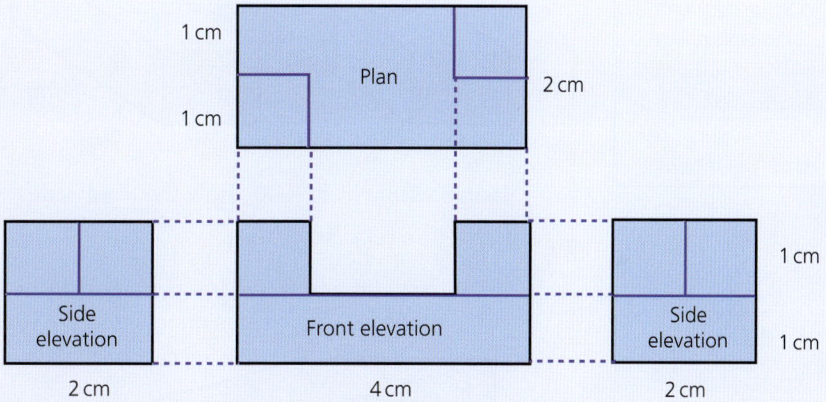

Side elevation

Front elevation

Side elevation

1 cm

1 cm

2 cm

4 cm

2 cm

SECTION 3

Exercise 23.1

1 The following shape is a cylindrical section of concrete tubing being used for the construction of a tunnel. The outer radius of the tube is 4.5 m while the inner radius is 3.75 m.

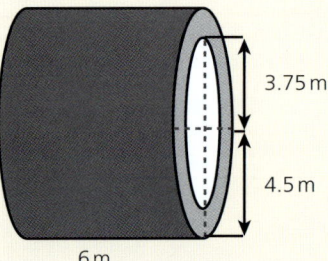

Using a scale of 1:150, draw the front, side and plan elevations of the tube.

2 A simplified 3D drawing of a new hotel is shown below.

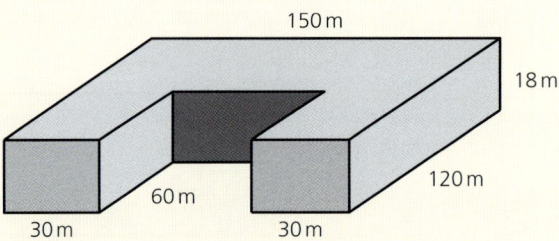

Squared paper or graph paper will help you draw these elevations more accurately than plain paper.

Using a scale of 1:3000, draw the front, side and plan elevations of the hotel.

3 A large warehouse for the distribution company is shown below. The main part of the building is a pentagonal-shaped prism. It also has a cube-shaped entrance area attached to the front of the building.

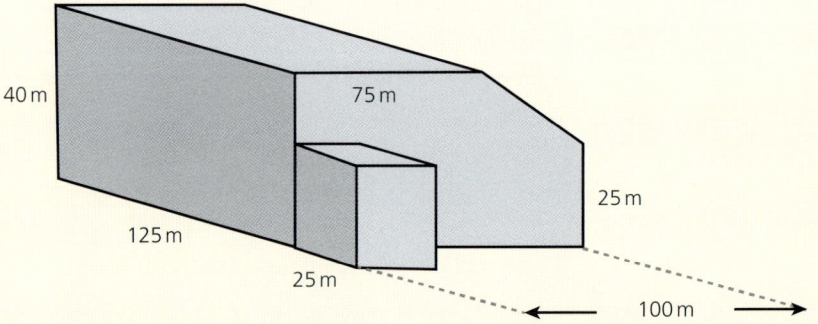

Using a scale of 1:2500, draw the front, side and plan elevations of the warehouse.

23 2D representations of 3D shapes

LET'S TALK
How does being asked to draw the elevations to a scale of 5:1 differ from a scale of 1:5?

4 A metal disc is cut so that a quarter of it is removed as shown:

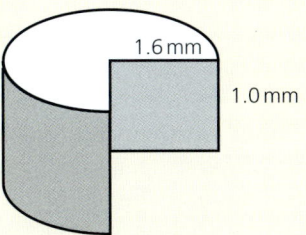

Using a scale of 10:1, draw the front, side and plan elevations of the three-quarter disc.

5 A plan of an object is shown below:

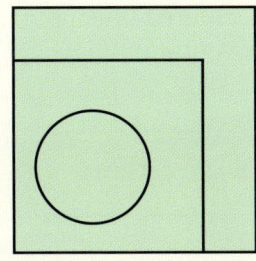

LET'S TALK
If the shape was rotated 30°, would the elevations change?

If so, *sketch* what they might look like.

a   Sketch two possible 3D objects that would have this plan view.
b   For one of your sketches, decide on its dimensions.
c   For the object you have chosen, choose an appropriate scale and draw its front, side and plan elevations.

Now you have completed Unit 23, you may like to try the Unit 23 online knowledge test if you are using the Boost eBook.

213

24 Functions

- Understand that a function is a relationship where each input has a single output.
- Generate outputs from a given function and identify inputs from a given output by considering inverse operations (including fractions).
- Understand that a situation can be represented either in words or as a linear function in two variables (of the form $y = mx + c$), and move between the two representations.

Function machines

In Stage 7, you looked at how function machines carry out mathematical operations. A number is entered (the input) and the function machine calculates an answer (the output).

For example:

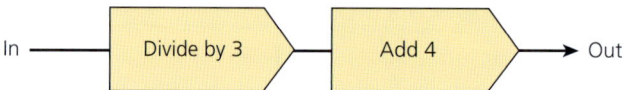

In this case, the function is 'divide by 3 and then add 4'.

An important property of a function is that one input value only produces one output value.

For example, the machine shown below is not an example of a function machine:

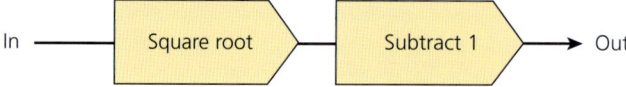

KEY INFORMATION
±3 means the answer can be +3 or −3

If the input is 9, then $\sqrt{9} = \pm 3$ because $(-3) \times (-3) = 9$ as well as $3 \times 3 = 9$.

24 Functions

> **Worked example**

a The numbers 3, 2, 1, 0, −1, −2 are entered into the function machine.

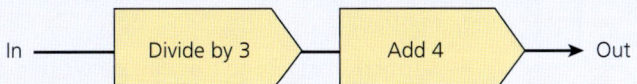

Calculate the output in each case.

Input	Output
3	5
2	$4\frac{2}{3}$
1	$4\frac{1}{3}$
0	4
−1	$3\frac{2}{3}$
−2	$3\frac{1}{3}$

The inverse function undoes the effect of the original function.

b A function machine is given as:

i) If the output is $\frac{5}{2}$, calculate the input.

The **inverse function** will help to solve this problem.

Working backwards, we must work out what mathematical operation undoes the original function. The opposite of 'division' is 'multiplication', while the opposite of 'subtraction' is 'addition'. The original function can therefore be reversed as follows:

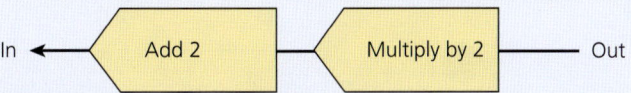

Therefore inserting $\frac{5}{2}$ as the output gives:

Therefore if $\frac{5}{2}$ is the output, 7 was the input.

215

SECTION 3

> The equation $y = 5x + 3$ can also be called an algebraic function.

c Express the equation $y = 5x + 3$ as a function machine.

Here, whatever the input x, it is multiplied by 5 and then 3 is added to it to produce the y-value. Therefore x is the input and y the output.

As a function machine this can be written as:

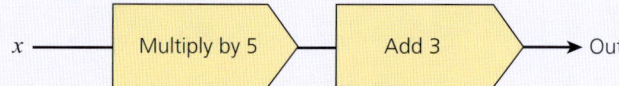

Exercise 24.1

1 Write each of the following functions in machine form.

$y = \frac{1}{2}x + 6$

$m = \frac{n+3}{2}$

In questions 2–4, the input numbers are listed in the table. Calculate the output values.

2

Input	Output
0	
1	
2	
3	

3

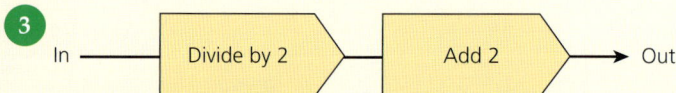

Input	Output
2	
1	
0	
−1	

24 Functions

4

Input	Output
−2	
−1	
0	
1	
2	

5 In the following question, a function machine is given:

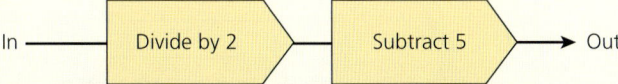

a Write down the inverse function machine.
b Complete the table of inputs and outputs.

Input	Output
4	
	18
25	
	307

6 Kimie and Fusayo are discussing the function machine below:

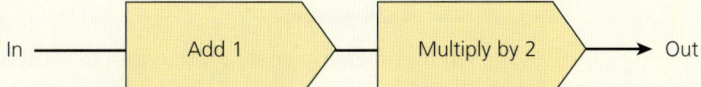

To rewrite it as an algebraic function, Kimie suggests the following:
$$x + 1 \times 2 = y$$
She says that because of BIDMAS the multiplication is done first so it can be written as: $x + 2 = y$.
Fusayo thinks it is a different algebraic rule.
a Prove, using numbers, that Kimie is wrong.
b If Fusayo's alternative is correct, write down the algebraic function that she wrote.

> To prove something is wrong you only need to find one example that it does not work.

SECTION 3

LET'S TALK
The machine functions for parts (i) and (ii) are different because they have been swapped around. However, are there any input values that will give the same output for each pair of machine functions?

7 Write each of the following function machines using algebra.

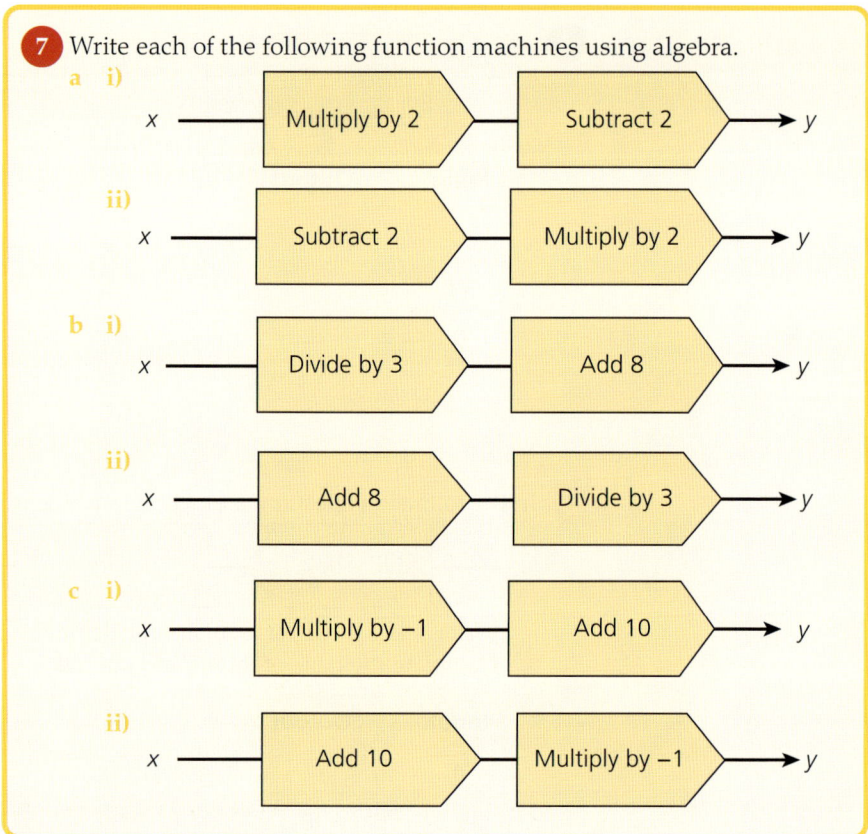

Creating a function machine

Worked example

A gym membership costs $150 to join and then an additional $80 per month.

i) Write, in function machine form, the relationship between the total cost (C) and the number of months (m) of membership.

ii) Write your function machine as an algebraic function.
$C = 80m + 150$

iii) A member worked out that she has paid a total of $7190 since she joined.
Calculate how long she has been a member.

24 Functions

To work this out the inverse function will be needed. This can be written as follows:

Giving the following solution:

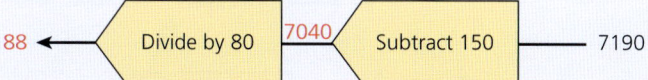

Therefore, she has been a member for 88 months, or 7 years and 4 months.

Exercise 24.2

1. A recipe states that to cook a turkey it takes 30 minutes per kilogram plus 20 minutes.
   a Construct a function machine to calculate the total time T in minutes needed to cook a turkey of m kg.
   b Write the function algebraically.
   c Calculate the time needed to cook a $4\frac{1}{2}$ kg turkey. Give your answer in hours and minutes.
   d Calculate the mass of a turkey that takes 4 hours and 20 minutes to cook.

2. It is known that in order to work out the sum (S) of the internal angles of an n-sided polygon, 2 is subtracted from the number of sides and the result multiplied by 180.
   a Draw a machine function linking S and n.
   b Write your machine function algebraically.
   c If a polygon has 30 sides, calculate the sum of its internal angles.
   d The sum of the internal angles of a polygon is 3600°. Calculate the number of sides of the polygon.
   e Are there values of n for which the function is not valid? Justify your answer.

3. A temperature in Celsius (C) can be converted to Fahrenheit (F) using the following function machine.

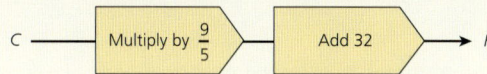

   a Write the rule as an algebraic function.
   b A student suggests that the inverse function machine can be written as:

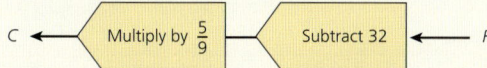

   Is this a possible inverse function? Justify your answer.

LET'S TALK

Is the algebraic function in part (b) an example of an equation or a formula?

SECTION 3

4 A candle maker makes candles that are 28 cm long. He states that the candles are long-lasting as they only burn by 1.5 cm per hour. Four possible function machines are shown below, where the length remaining (L) of the candle is given in terms of the number of hours (n) it has been lit.

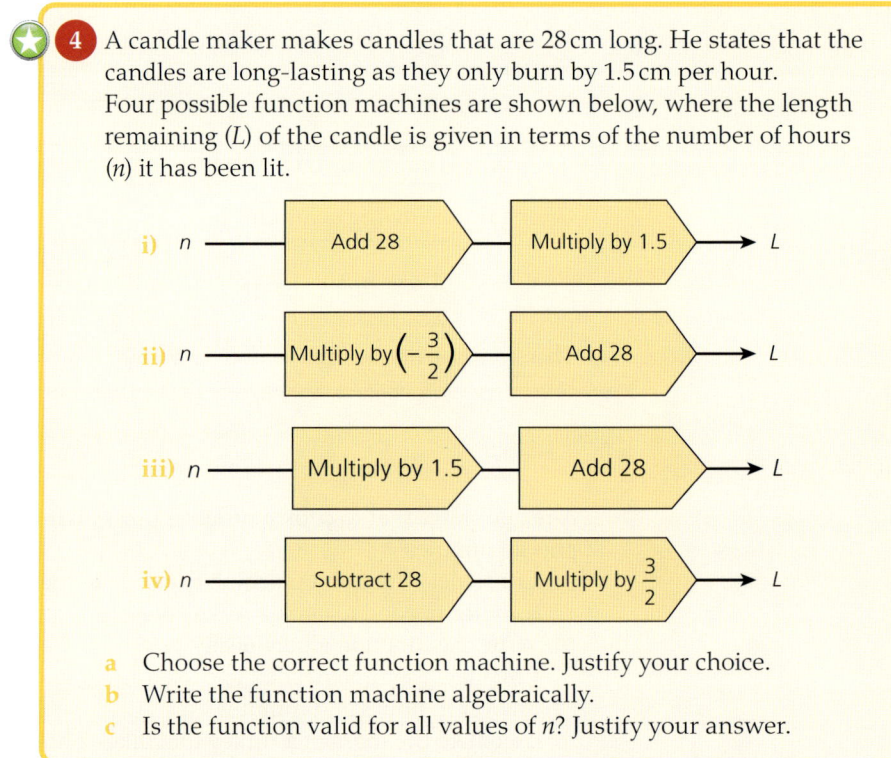

a Choose the correct function machine. Justify your choice.
b Write the function machine algebraically.
c Is the function valid for all values of n? Justify your answer.

Now you have completed Unit 24, you may like to try the Unit 24 online knowledge test if you are using the Boost eBook.

Geometry and translations

- Use knowledge of coordinates to find the midpoint of a line segment.
- Translate points and 2D shapes using vectors, recognising that the image is congruent to the object after a translation.

The midpoint of a line segment

A **line segment** is part of a line and has a fixed length.

This is in contrast to a **line**, which is of infinite length.

The midpoint of a line segment is the point exactly halfway along its length.

You already know how to find a midpoint of a line by constructing a perpendicular bisector as shown in Unit 17.

This unit will look how to find the midpoint using the coordinates of the end points of the line segment.

If the coordinates of the two end points of a line segment are known, it is possible to calculate the coordinates of its midpoint and also the length of the line segment.

LET'S TALK
Is it possible to draw a true line?

Worked example

Two points A = (3, 1) and B = (9, 5) are joined by the straight-line segment AB.

Calculate the coordinates of the midpoint, M, of AB.

The diagram shows points A and B and the line segment joining them, plotted on a coordinate grid.

The horizontal distance between A and B is 9 − 3 = 6 units.

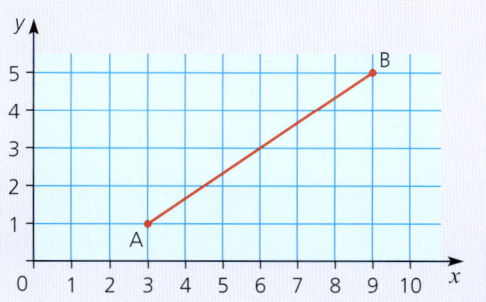

SECTION 3

The vertical distance between A and B is
5 − 1 = 4 units.

The midpoint is at the point which is half of each of these distances away from A and B. That is, it is 3 units horizontally from A (and B) and 2 units vertically from A (and B).

Therefore, the coordinates of the midpoint M from A are (3 + 3, 1 + 2) = (6, 3).

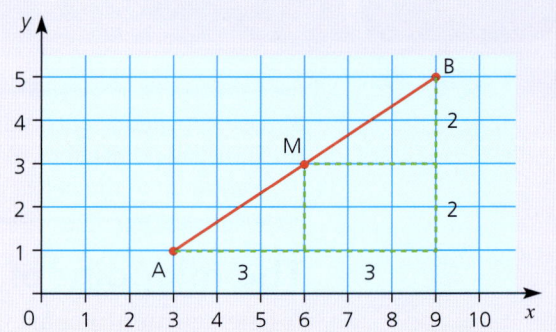

LET'S TALK
How can the midpoint be calculated from B?

The calculation can be simplified as follows. The x- and y-coordinates of the midpoint of a line segment are the means of the x- and y-coordinates of the end points of the line segment. In the example above, the coordinates of M are $\left(\frac{3+9}{2}, \frac{1+5}{2}\right) = (6, 3)$.

In **general**:

If the coordinates of the two end points of a line segment are (x_1, y_1) and (x_2, y_2), the coordinates of the midpoint are $\left(\frac{x_1+x_2}{2}\right), \left(\frac{y_1+y_2}{2}\right)$.

Exercise 25.1

1. For each of the following pairs of points:
   i) plot the points on a coordinate grid and join them with a line segment
   ii) calculate the coordinates of the midpoint of the line segment.
   a A = (1, 3) and B = (7, 5)
   b X = (2, 1) and Y = (4, 7)
   c P = (−5, 1) and Q = (6, 1)
   d L = (−6, 4) and M = (−6, 9)
   e J = (−2, −3) and K = (−5, 2)
   f C = (−1, 8) and D = (2, −3)
2. A line segment AB has midpoint M = (3, 5) and the coordinates of point A are (7, 3). Calculate the coordinates of point B.
3. A line segment ST has midpoint M = (−2, 4) and the coordinates of point T are (1, 9). Calculate the coordinates of point S.

25 Geometry and translations

 4 The diagram shows a square ABCD.
The diagonals AC and BD intersect at their midpoint M = (0, −2).
Calculate the coordinates of the vertices A, B and D.

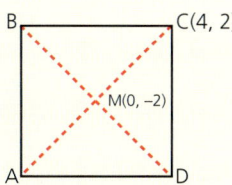

 5 The diagram shows a parallelogram DEFG.
The point M is the midpoint of the diagonals DF and EG. Calculate
   a the coordinates of M
   b the coordinates of the vertex D.

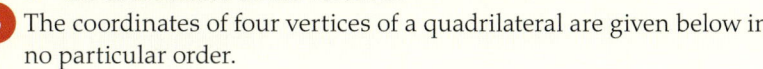

6 The coordinates of four vertices of a quadrilateral are given below in no particular order.

(−1, 2)   (2, −1)

(3, 8)   (6, 5)

The midpoint of the line segments joining opposite vertices is the same point.
Deduce the coordinates of the midpoint.

Translation

Earlier, in Unit 13, you looked at some of the different forms of transformation.

These included reflections, rotations and enlargements. With two of these the object and image are congruent.

LET'S TALK
Which two produce an image congruent to the object?

A fourth transformation, covered briefly in Stage 7, are **translations**.

A translation refers to a straight sliding motion without rotation.

In diagram below, triangle ABC (the object) has been translated to its new position A'B'C' (the image).

LET'S TALK
Will a translation always produce a congruent shape?

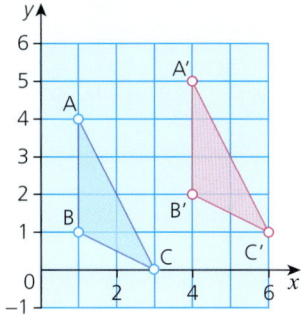

SECTION 3

Each of the vertices of ABC have been moved three units across to the right and one unit up.

In a translation, every point on the object moves by exactly the same amount.

Rather than describing the translation as 3 across to the right and 1 up, it is more usual to use a **column vector**.

In this example, the vector is written as $\begin{pmatrix} 3 \\ 1 \end{pmatrix}$.

The top number indicates how far the object moves horizontally. If it is a positive number it is a move to the right, while a negative number implies a move to the left.

The bottom number indicates how far the object moves vertically. If it is a positive number it is a move upwards, while a negative number implies a move downwards.

Worked example

The grid below shows an object PQRS translated to a new position P'Q'R'S'.

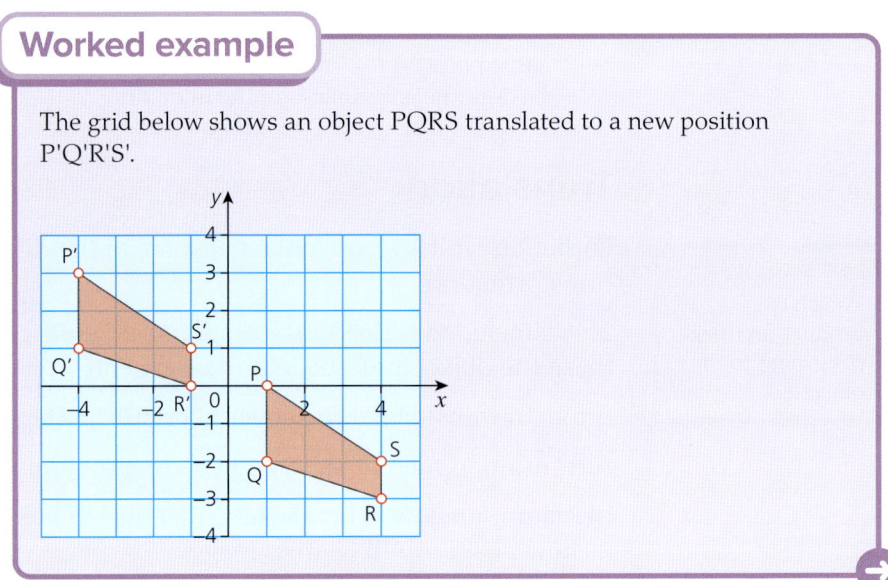

25 Geometry and translations

Work out the column vector.

In order to visualise the translation, a **vector** can be drawn to connect two corresponding vertices and also to show the direction of travel.

The vector is a straight line, here joining Q to Q', with a small arrowhead indicating the direction.

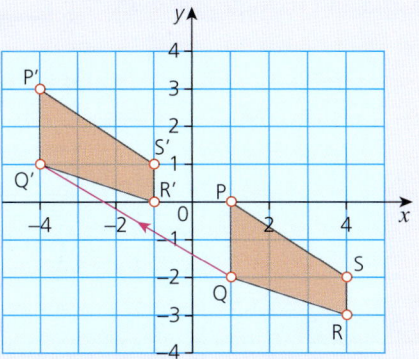

Only one vector line has been drawn here, but a vector line joining P to P', or S to S' etc. would be exactly the same size and direction as the one drawn.

We can see that the vector line shows a translation of 5 units to the left and 3 units up.

As a column vector, this is written as $\begin{pmatrix} -5 \\ 3 \end{pmatrix}$.

Note that the top number is negative, indicating a move to the left, and the bottom number is positive, meaning a move upwards.

Exercise 25.2

1 A parallelogram PQRS is translated to P'Q'R'S'. P'Q'R'S' is then translated to P"Q"R"S".

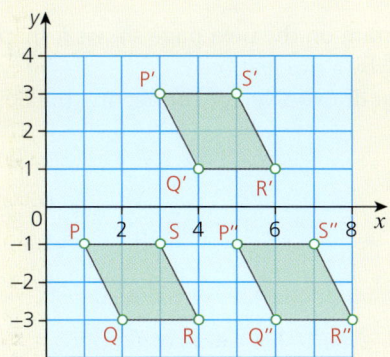

225

SECTION 3

a Write the column vector that maps PQRS to P'Q'R'S'.
b Write the column vector that maps P'Q'R'S' to P"Q"R"S".
c Write the column vector that maps PQRS to P"Q"R"S".

LET'S TALK
Is there a connection between your answers to parts (a), (b) and (c)?
Can you explain any findings?

2 Copy the grid and triangle A shown below.

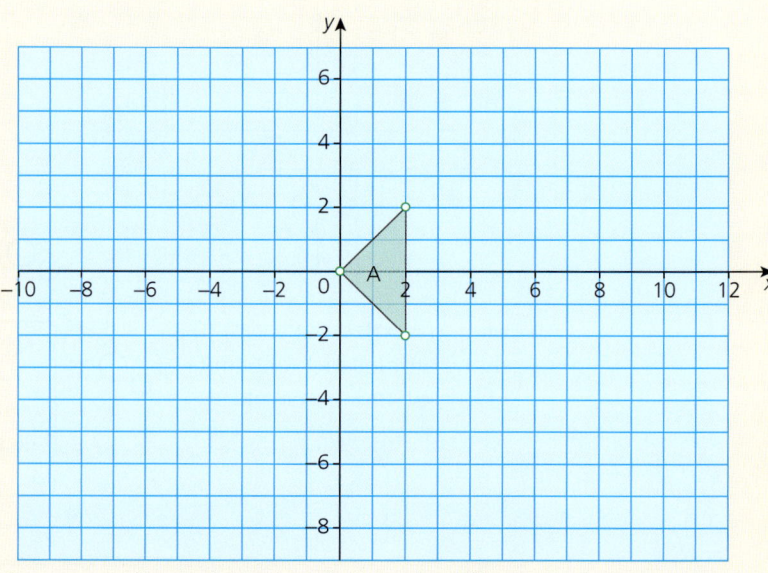

On your grid, translate triangle A by the following column vectors:

a $\begin{pmatrix} 5 \\ 4 \end{pmatrix}$ and label it B

b $\begin{pmatrix} -6 \\ 1 \end{pmatrix}$ and label it C

c $\begin{pmatrix} 0 \\ -5 \end{pmatrix}$ and label it D

d $\begin{pmatrix} 8 \\ -2 \end{pmatrix}$ and label it E

3 The diagram on the next page shows a shape WXYZ mapped onto W'X'Y'Z' by a translation.
A student draws a vector connecting the object and its image as shown.

25 Geometry and translations

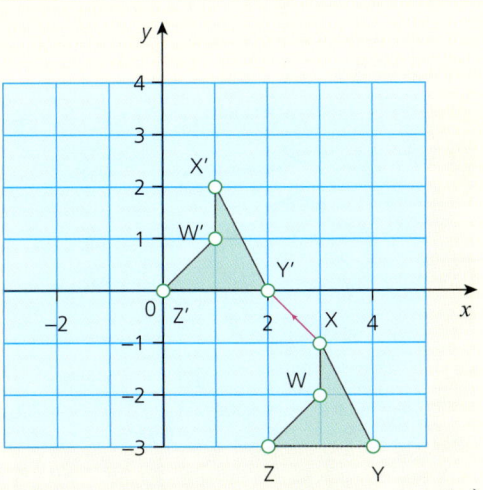

They state that the vector translation is $\begin{pmatrix} -1 \\ 1 \end{pmatrix}$.

If you agree with the statement, explain why. If you disagree with it, explain why and state an alternative column vector.

4 The grid below shows three congruent shapes, A, B and C.

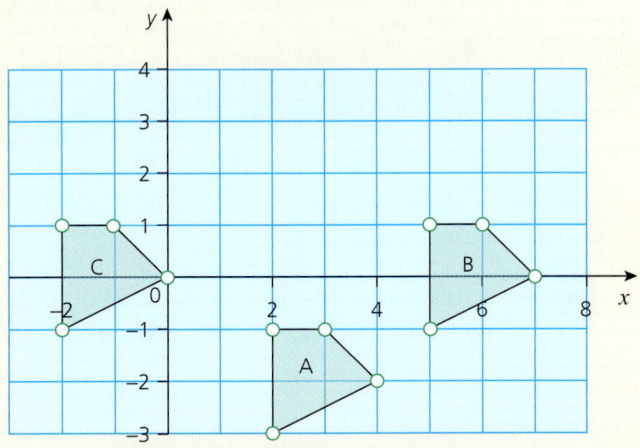

a   i) Write the vector that maps A onto B.
   ii) Write the vector that maps B onto A.
b   i) Write the vector that maps A onto C.
   ii) Write the vector that maps C onto A.
c   i) Write the vector that maps B onto C.
   ii) Write the vector that maps C onto B.

d If the vector that maps a shape X onto Y is given as $\begin{pmatrix} a \\ b \end{pmatrix}$, write the vector that maps Y onto X. Justify your answer.

SECTION 3

LET'S TALK

1. How many different translation vectors could this wallpaper pattern have?
2. Why do wallpaper designs have repeating patterns?

5 Most wallpaper designs have a repeating pattern involving translations.

a Design a simple wallpaper pattern of your own.
b Describe the different translations that appear on your pattern.

Now you have completed Unit 25, you may like to try the Unit 25 online knowledge test if you are using the Boost eBook.

26 Squares, square roots, cubes and cube roots

- Recognise squares of negative and positive numbers, and the corresponding square roots.
- Recognise positive and negative cube numbers, and the corresponding cube roots.

Squares and square roots

The numbers 1, 4, 9, 16, 25, 36, 49, 64, 81, 100, ... are **square numbers**, and are made by multiplying an integer by itself.

For example:

$7 \times 7 = 49$ and $8 \times 8 = 64$

Therefore, 49 and 64 are square numbers.

But $7.3 \times 7.3 = 53.29$

53.29 is not a square number as 7.3 is not an integer.

Squaring a number is multiplying a number by itself. For example,

8 squared is $8 \times 8 = 64$

7.3 squared is $7.3 \times 7.3 = 53.29$

Using indices, 8 squared is written as 8^2 and 7.3 squared is written as 7.3^2.

The inverse operation of squaring is finding the square root. For example,

You already know that if two negative numbers are multiplied together, the answer is positive.

For example: $(-8) \times (-8) = 64$.

SECTION 3

LET'S TALK
Is it possible to have the square root of a negative number? Justify your answer.

This therefore produces an interesting situation.

If $8^2 = 64$ and $(-8)^2 = 64$, then $\sqrt{64}$ can be equal to either $+8$ or -8.

This is written more commonly as $\sqrt{64} = \pm 8$

The square root of a positive number gives two answers, one positive and one negative.

Worked example

a i) Calculate $(-6)^2$
$(-6)^2 = (-6) \times (-6) = 36$

ii) Calculate $\sqrt{36}$
$6 \times 6 = 36$ and $(-6) \times (-6) = 36$
Therefore $\sqrt{36} = \pm 6$

b Without a calculator work out $\sqrt{0.16}$, giving your answer as a decimal.
Converting to a fraction first makes the calculation easier to visualise.
$\sqrt{0.16}$ can be written as $\sqrt{\frac{16}{100}} = \frac{\sqrt{16}}{\sqrt{100}} = \pm\frac{4}{10}$
Therefore $\sqrt{0.16} = \pm 0.4$

Note that the square root of a number between 0 and 1 will produce a bigger answer.

Exercise 26.1

Work out the square roots in parts (a) and (b) in questions 1–7.

1. a $\sqrt{25}$ b $\sqrt{36}$
2. a $\sqrt{49}$ b $\sqrt{64}$
3. a $\sqrt{121}$ b $\sqrt{144}$
4. a $\sqrt{81}$ b $\sqrt{100}$
5. a $\sqrt{0.01}$ b $\sqrt{0.04}$
6. a $\sqrt{0.36}$ b $\sqrt{0.49}$
7. a $\sqrt{0.25}$ b $\sqrt{0.16}$
8. Square the following numbers.
   a 11^2 b $(-12)^2$ c $(-15)^2$
   d $(-1)^2$ e $(0.5)^2$ f $(-0.25)^2$

26 Squares, square roots, cubes and cube roots

9 A square with an area of 169 cm² is shown below. It has a side length of x cm.

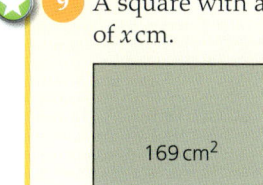

A student states that in order to find the length of x, you simply square root the area. He calculates the value of x to give $\sqrt{169} = \pm 12$.
He has made two errors. What are they? Justify your answers.

10 A sequence of squares makes a pattern. The first square has an area of 16 cm², the second an area of 8 cm², the third 4 cm² and so on, each time the area of the next square in the sequence being half that of the previous one. The first six squares in the pattern are shown:

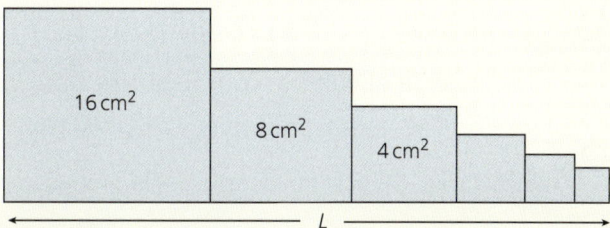

a Estimate the length marked L. Show your working.
b If a seventh square was added to the pattern, how much would the length L increase by?

Cubes and cube roots

This pattern sequence is made up of 1 cm cubes.

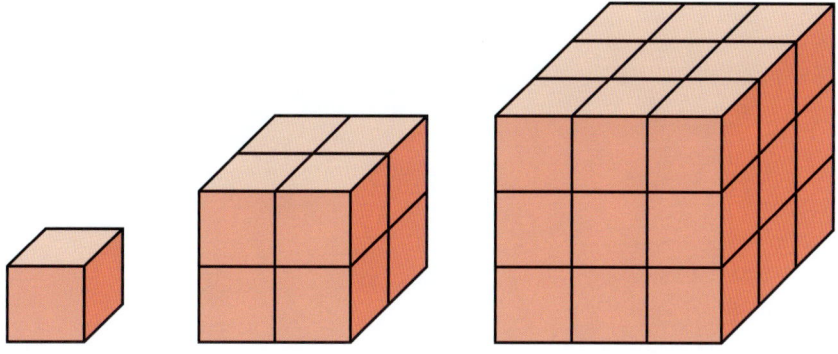

231

SECTION 3

The 1 cm × 1 cm × 1 cm cube contains *one* 1 cm × 1 cm × 1 cm cube.

The 2 cm × 2 cm × 2 cm cube contains *eight* 1 cm × 1 cm × 1 cm cubes.

The 3 cm × 3 cm × 3 cm cube contains *twenty-seven* 1 cm × 1 cm × 1 cm cubes.

The numbers 1, 8, 27, 64, 125, 216, 343, 512, 729, 1000, … are **cube numbers**, and are made by multiplying an integer by itself three times. For example, $5 \times 5 \times 5 = 125$.

Therefore 125 is a cube number.

Cubing a number is multiplying a number by itself three times.

As with squaring, there is a short way to write a number cubed using indices.

For example,

$5 \times 5 \times 5 = 5^3$ and $(-5) \times (-5) \times (-5) = (-5)^3$.

The inverse of cubing a number is finding its cube root, written as $\sqrt[3]{}$.

So $\sqrt[3]{125}$ is 5 and $\sqrt[3]{343}$ is 7 (since $7 \times 7 \times 7 = 343$).

Similarly, $\sqrt[3]{-125}$ is −5 and $\sqrt[3]{-343}$ is −7.

> The cubing of a negative number produces a negative answer.

> Although square rooting a negative number does not give a real answer, cube rooting a negative number does. i.e. $\sqrt[3]{-125} = -5$ because $(-5)^3 = -125$.

Worked example

Calculate $\sqrt[3]{(6)^3}$.

Cubing and cube rooting are inverse operations (i.e. they undo the effect of each other) therefore $\sqrt[3]{(6)^3} = 6$.

The above answer can be visualised as follows: $\sqrt[3]{(6)^3} = \sqrt[3]{6 \times 6 \times 6} = 6$.

26 Squares, square roots, cubes and cube roots

Exercise 26.2

Work out the cube roots in parts (a) and (b) in questions 1–4.

1. a $\sqrt[3]{64}$ b $\sqrt[3]{125}$
2. a $\sqrt[3]{343}$ b $\sqrt[3]{-512}$
3. a $\sqrt[3]{216}$ b $\sqrt[3]{-729}$
4. a $\sqrt[3]{8000}$ b $\sqrt[3]{-1728}$
5. a Explain why $7^2 \times 7 = 7^3$.
   b Explain why $\left(\sqrt[3]{64}\right)^3 = \sqrt[3]{64^3} = 64$.
6. A student states that $\sqrt[3]{-216} = \pm 6$.
   State whether the answer is correct and justify your answer.
7. a Show that a pile of seventy-two $1 \times 1 \times 1$ cm³ cubes can be rearranged to form two larger cubes.
   b i) What is the smallest number of cubes that can be made from a pile of forty-three $1 \times 1 \times 1$ cm³ cubes?
   ii) Justify your answer.

Now you have completed Unit 26, you may like to try the Unit 26 online knowledge test if you are using the Boost eBook.

Graphs and equations of straight lines

- Use knowledge of coordinate pairs to construct tables of values and plot the graphs of linear functions, where y is given explicitly in terms of x ($y = mx + c$).
- Recognise that equations of the form $y = mx + c$ correspond to straight-line graphs, where m is the gradient and c is the y-intercept (integer values of m).

> As its name implies, the value of a variable can change.

Coordinate pairs

You will already be familiar with formulae linking two variables.

For example, with the equation $y = 2x - 4$, y is given in terms of x, so that as the value of x changes then the value of y will too.

The values for x and y can then be plotted as coordinates on a graph.

Worked example

a i) Complete the table of values below for the equation $y = 2x - 4$.

x	y
10	
5	
0	
−3	

→

x	y
10	16
5	6
0	−4
−3	−10

ii) Plot the x and y values as coordinates and draw a line through them.

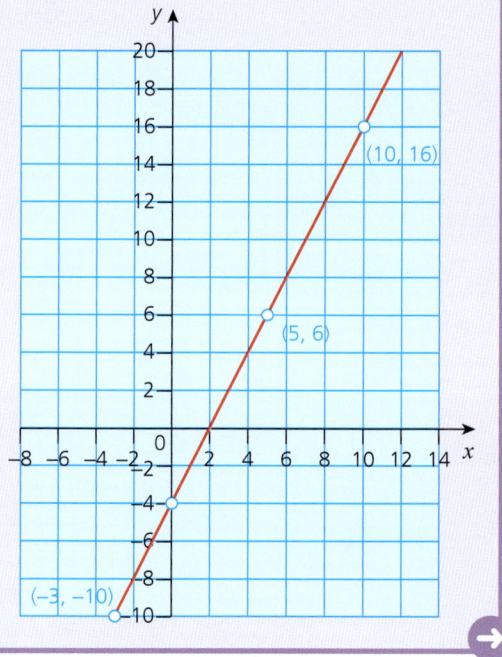

234

27 Graphs and equations of straight lines

iii) Does the point with coordinates (15, 26) lie on the line? Justify your answer.

Yes, the point does lie on the line as the coordinates (15, 26) fit the equation.

i.e. $y = 2x - 4$

$26 = 2 \times 15 - 4$

$26 = 26$

A straight line when plotted on a graph will usually cross both the x-axis and the y-axis.

We know that whenever a line crosses the x-axis, the y-coordinate will be zero. Similarly, where the line crosses the y-axis, the x-coordinate will be zero.

If a point lies on a line, then its coordinates fit the equation of the line.

LET'S TALK
What type of straight lines will cross the axes only once?

Worked example

a i) Work out where the line produced by the equation $y = 3x + 6$ crosses the x- and y-axes.

When it crosses the y-axis, $x = 0$. Substituting this into the equation gives:

$y = 3 \times 0 + 6$

$y = 6$ Therefore, it crosses the y-axis at (0, 6)

When it crosses the x-axis, $y = 0$. Substituting this into the equation gives:

$0 = 3x + 6$

This must be rearranged to find x:

$-6 = 3x$ (subtract 6 from both sides)

$-2 = x$ (divide both sides by 3)

Therefore, it crosses the x-axis at (−2, 0)

Although in theory only two points are needed to be able to draw a straight line, it is good practice to plot a third point just to check that you have not made a mistake. All three points should form a straight line.

ii) Plot the line produced by the function $y = 3x + 6$.

As we already know that it passes through the points (0, 6) and (−2, 0), these can be plotted and a line drawn through them.

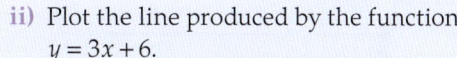

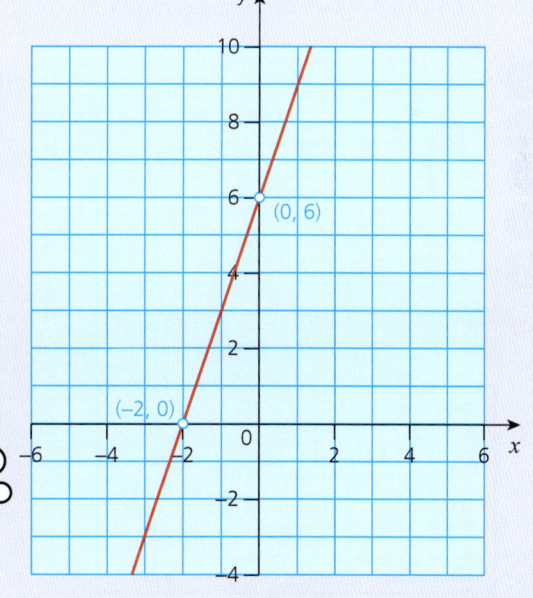

SECTION 3

b Draw the line $y = 2x - 1$ on a coordinate grid.

To find the positions of two points on the line, choose two values for x. Substitute each of these into the equation and calculate the y-value.

When $x = 0$, $y = -1$, giving the coordinates $(0, -1)$.

When $x = 3$, $y = 5$, giving the coordinates $(3, 5)$.

Plotting these two points and drawing the line between them gives the following graph.

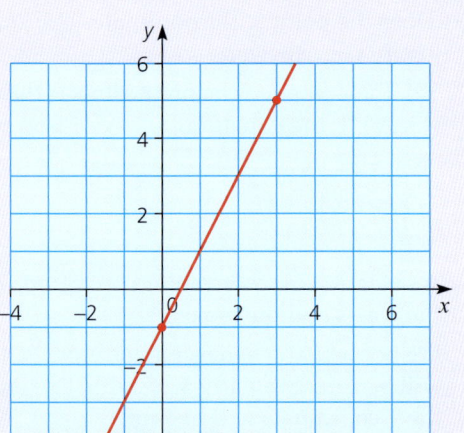

> It is good practice to check a third point.
>
> When $x = 1$, $y = 1$.
>
> As the point $(1, 1)$ lies on the line, the graph is correct.

Exercise 27.1

On a coordinate grid, draw the straight line represented by each of these equations.

> First identify the coordinates of three points on the line.

1. $y = x + 2$
2. $y = 2x - 3$
3. $y = \frac{1}{2}x + 1$
4. $y = 3$
5. $y = x - 1$
6. $x = -2$
7. $y = \frac{1}{2}x + 3$
8. $y = -x + 3$
9. $y = -2x + 2$
10. $y = -x - 1$

11. The following Venn diagram represents the equations of two straight lines, $y = 3x - 5$ and $y = 2x + 1$. Next to it are the coordinates of six points.

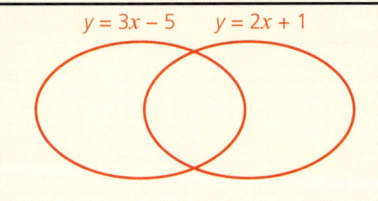

(1, 3)   (6, 13)

(4, 5)   (−2, −3)

(0, −1) (−1, −8)

236

27 Graphs and equations of straight lines

a Copy the Venn diagram and place the coordinates in the correct region.
b What is the maximum number of points that can be placed in the overlap of the Venn diagram? Justify your answer.

12 The coordinates of four points are given below.
(5, 1)   (0, 5)   (3, 2)   (−2, 7)
Three of the points lie on the same line.
a Which point does not lie on the same line as the others? Justify your answer.
b If the x-coordinate of this point is correct, calculate the correct y-coordinate.

13 A computer repair specialist charges a flat rate of $30 to look at a faulty computer and then a further $18 per hour to repair it. She only charges for whole hours worked.
Three customers, A, B and C, take their computers to be looked at and repaired.
Their bills are as follows:

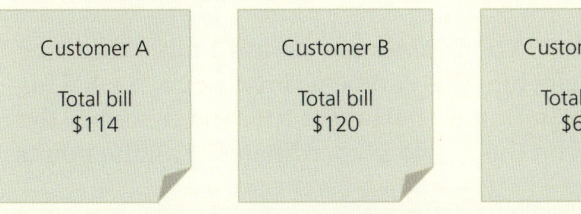

Customer A — Total bill $114
Customer B — Total bill $120
Customer C — Total bill $66

One of the customers complains that his bill must be incorrect. Which customer complains? Justify your answer.

Linear graphs

A line is made up of an infinite number of points. As shown above, only two of those points are needed in order to be able to draw a line.

For example:

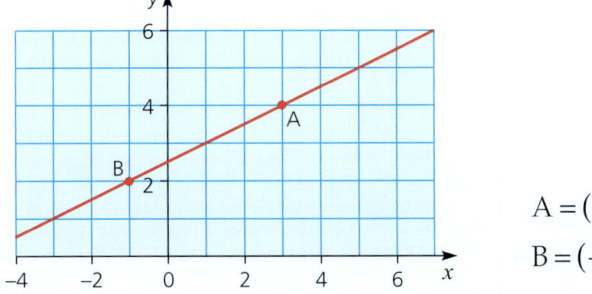

$A = (3, 4)$
$B = (-1, 2)$

The coordinates of every point on a straight line all have a common relationship, that is, there is a rule which the x- and y-coordinates all follow.

SECTION 3

The line below is plotted on a coordinate grid.

By putting the coordinates of some of the points in a table, we can see a pattern linking the x- and y-values.

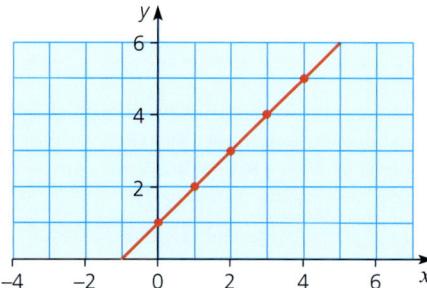

x	y
0	1
1	2
2	3
3	4
4	5

We can see that the y-coordinates are always 1 more than the x-coordinates. In algebra, this can be written as $y = x + 1$. This is known as the **equation of a straight line** and describes the relationship between the x- and y-coordinates of all the points on the line.

Exercise 27.2

For each of the straight lines shown in questions 1–5, write in a table the coordinates of five of the points on the line and deduce the equation of the line.

1

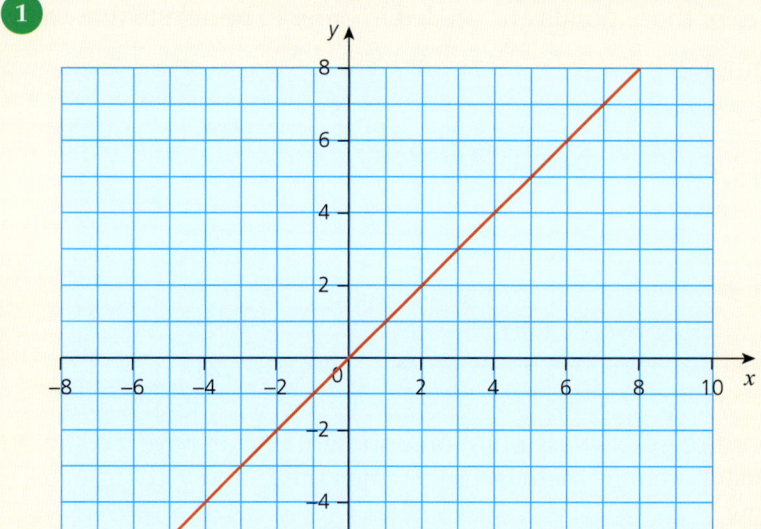

27 Graphs and equations of straight lines

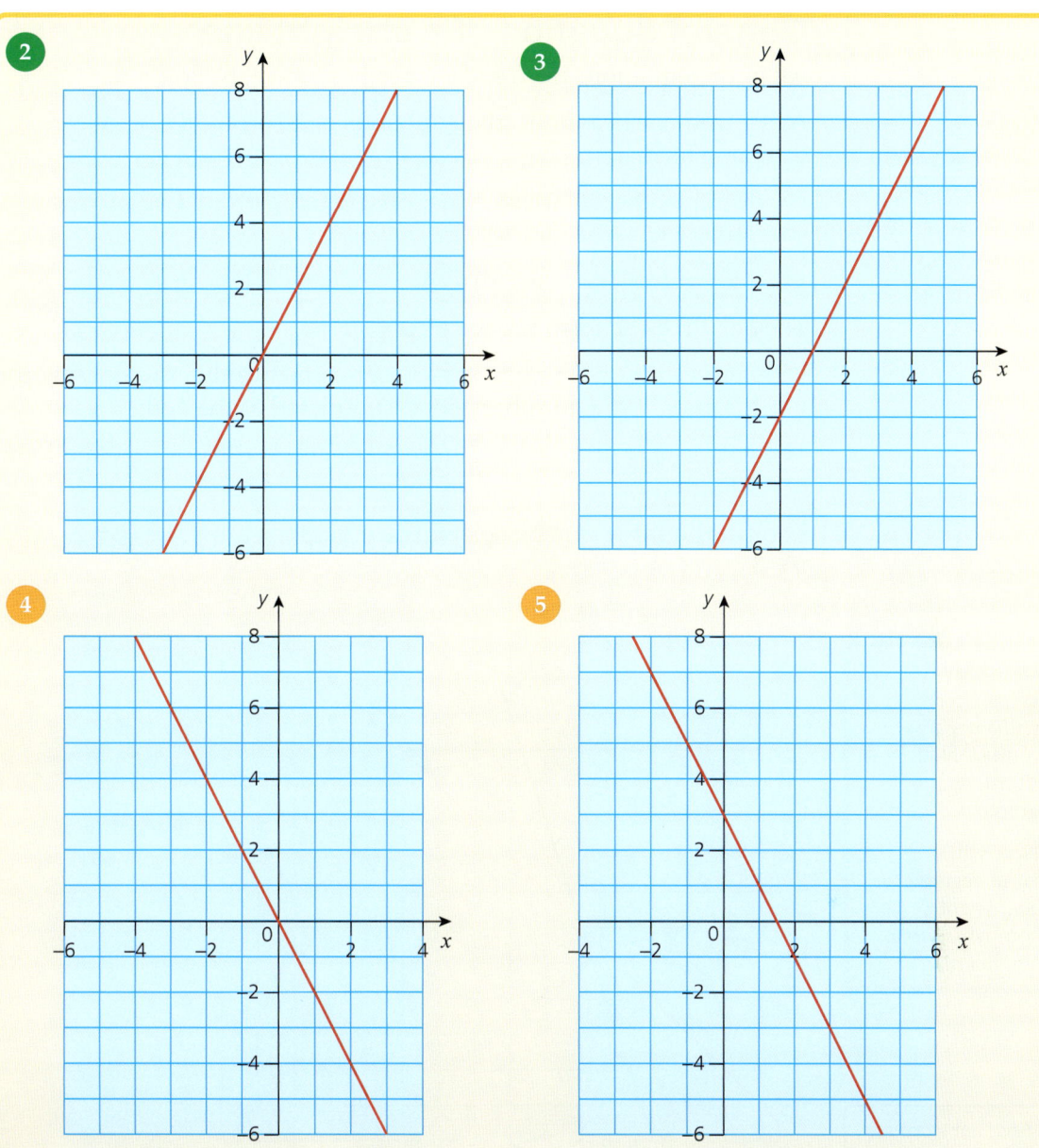

6. By looking at each of the following equations, decide whether, if points were plotted, they would produce a horizontal, a vertical or a sloping line.

a $y = x$
b $y = x + 4$
c $x = 7$
d $y = -x$
e $y = 6$
f $x = -2$
g $y = 4x + 6$
h $y = -5x - 2$

LET'S TALK

How can you decide just by looking at the equation of a line whether it will be horizontal, vertical or sloping?

SECTION 3

> The form $y = mx + c$ is known as the **general form** of a straight line.

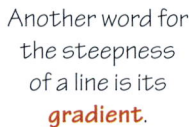

General equation of a straight line

All of the equations used so far have produced straight lines.

Apart from vertical lines, the equations of all straight lines have the same format.

For example, the following equations, when plotted, will produce straight lines.

$$y = 2x - 5 \qquad y = -3x + 1 \qquad y = x + 10$$

They all take the form known as $y = mx + c$.

When y is the subject, then m represents the number multiplying the x-value and c is the number added/subtracted at the end.

The values of m and c have different effects on the position and look of a straight line.

Look at the three graphs below. The equations are $y = x$, $y = 2x$ and $y = 3x$.

> **LET'S TALK**
>
> What effect will a negative coefficient of x have on the graph?
>
> How will a large negative coefficient compare with a smaller one?

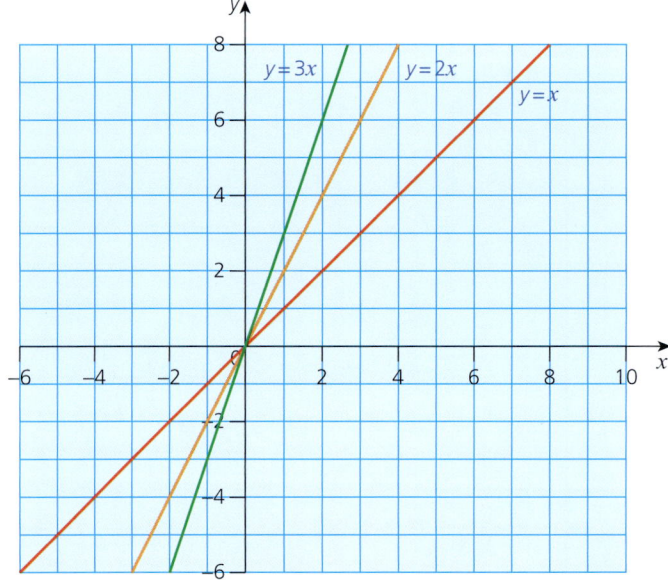

> Another word for the steepness of a line is its **gradient**.

We can see that the only change in each of the equations is the number multiplying the x (called the coefficient of x). The difference between each of the graphs is the steepness of each line. Therefore, it can be concluded that the coefficient of x indicates the steepness of the line. The larger the number, the steeper the line.

Look at three more graphs below.

Their equations are $y = 2x - 2$, $y = 2x + 1$ and $y = 2x + 4$.

As the x-coefficients are the same, the lines are parallel.

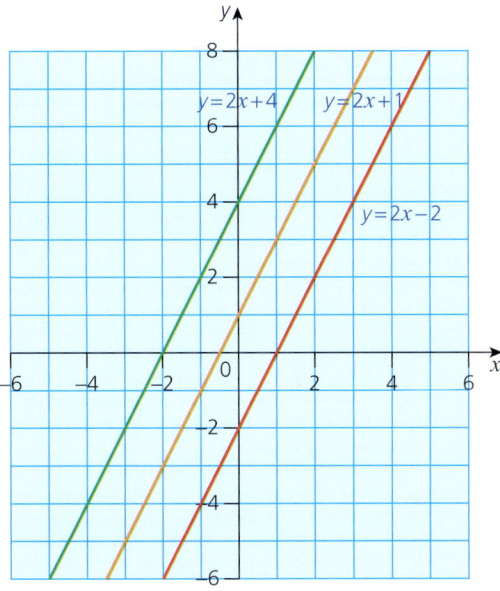

The coefficients of x are the same in each equation; therefore, their steepness is the same.

What is different about each equation is the number added/subtracted at the end. What is different about each of the graphs is where the lines **intersect** the axes. Looking closely, we can see that the number added/subtracted at the end is the same as where the line intersects the y-axis.

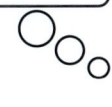

If m is positive, the straight line will slope this way:

If m is negative, the straight line will slope this way:

There is a reason for this. When the line intersects the y-axis, $x = 0$.

If this is substituted into the general equation $y = mx + c$, we get:

$$y = m \times 0 + c$$
$$y = c$$

So, when $x = 0$, the line intersects the y-axis and at that point $y = c$.

Summarising the findings:

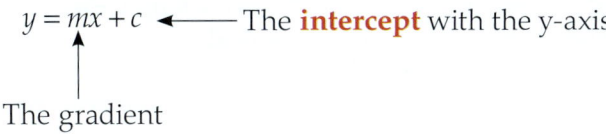

$y = mx + c$ ⟵ The **intercept** with the y-axis

The gradient

SECTION 3

Worked example

The graph below shows three lines, A, B and C. The numbers on the axes have been left blank. One line has equation $y = 2x - 5$, another has equation $y = -2x + 2$ and the third has equation $y = x - 5$.

Match each equation to each line. Justify your answers.

With $y = -2x + 2$, the coefficient of x is negative, so the straight line must slope downwards from left to right, i.e. therefore, it must be line A.

Lines B and C must therefore be $y = 2x - 5$ and $y = x - 5$.

Both have the same y-intercept as shown in the graph as they cross the y-axis at the same point. This must be at -5 as both equations end with -5.

As line B is steeper than line C, the coefficient of x in the equation for line B must be bigger. Therefore, line B is $y = 2x - 5$ and line C is $y = x - 5$.

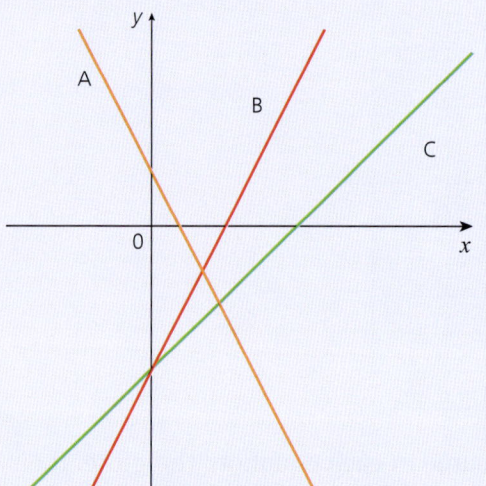

LET'S TALK

How can we have concluded that $y = -2x + 2$ is line A by looking at the y-intercepts?

Exercise 27.3

 1 The equations of two straight lines are given as follows: $y = 6x - 4$ and $y = 3x + 1$.
   a   Write an equation for a straight line with a gradient in-between that of the two lines given.
   b   Write an equation for a straight line with a y-intercept in-between that of the two lines given.

 2 The equations of six straight lines are given below.
   $y = 3x + 4$ $y = 2x + 5$ $y = 3x - 5$
   $y = -3x + 5$ $y = x + 5$ $y = -2x + 4$
   a   Which of the lines are parallel to each other?
   b   Which of the lines intersect the y-axis at the same point?
   c   Write the equation of a line that is parallel to $y = x + 5$ and intersects the y-axis at the same place as the line $y = -2x + 4$.

 3 The **general** equation of a straight line takes the form $y = mx + c$.
   a   Explain how the line $y = 3x$ fits the **general** form.
   b   i) Describe the steepness of the line $y = 6$.
       ii) Explain how the line with equation $y = 6$ fits the **general** form.

27 Graphs and equations of straight lines

  4 The equations of seven lines are listed below:
$y = -x$ $y = 2x + 3$ $y = -1$
$x = -4$ $y = -x + 3$ $y = x + 3$ $x = 4$

These seven lines and one other are graphed below.
  a Match each line to each equation.
  b i) The equation of which line does not appear in the list above?
     ii) Write a possible equation for this line. Justify your answer.

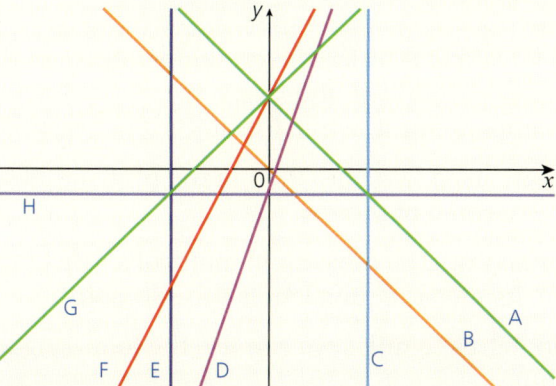

  5 An equilateral triangle and rectangle are shown below:

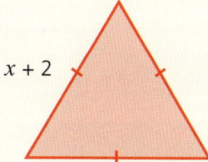   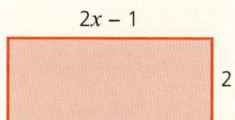

LET'S TALK
What is the smallest value of x that is valid for *both* perimeters?

  a Write a formula for the perimeter (y) of the equilateral triangle.
  b Write a formula for the perimeter (y) of the rectangle.
  c Plot a graph on the same axes, showing how the perimeter of each shapes changes with x.
  d From the graph, how can you deduce the value of x that gives both shapes the same perimeter? Justify your answer.

  6 A mobile phone company has two monthly tariffs for its latest phone. A user can opt for tariff P which is to pay a flat rate of $40 per month with unlimited data, or choose tariff Q which is to pay a rate of $25 per month with an additional cost of $4 per GB of data used.
  a Write a formula for the monthly cost (C) of using (G) GB of data under each of the tariffs.
  b i) If the cost of both tariffs over time was graphed, which tariff would have a steeper line? Justify your answer.
     ii) Which tariff would intersect the y-axis at a higher point? Justify your answer.

  Now you have completed Unit 27, you may like to try the Unit 27 online knowledge test if you are using the Boost eBook.

Distances and bearings

- Know that distances can be measured in miles or kilometres, and that a kilometre is approximately $\frac{5}{8}$ of a mile or a mile is 1.6 kilometres.
- Understand and use bearings as a measure of direction.

Distances

In Unit 6, you looked at how to convert temperatures in degrees Celsius to degrees Fahrenheit and vice versa. Fahrenheit is an example of an imperial unit, while Celsius is a metric unit.

LET'S TALK
1. Why do metric units tend to be easier to use than imperial units?
2. What other imperial units do you know? How do they compare to metric units?

Imperial units are used less now than in the past, although some countries still use some of them.

Another common imperial measurement involves distance and is called the mile. Its metric equivalent is the kilometre.

To convert kilometres to miles we can use the following approximate function machine

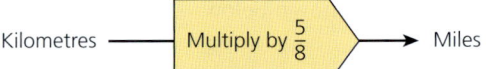

Worked example

During a car journey, the driver notices that he still has 240 km to travel. How far is this in miles?

Multiply the distance in kilometres by $\frac{5}{8}$ to convert it to miles.

$240 \times \frac{5}{8} = 150$, therefore, he still has 150 miles left to travel.

28 Distances and bearings

Exercise 28.1

1. Write the following distances in ascending order.
   135 miles   200 miles   208 km   327 km

2. The function machine to convert kilometres to miles is given as:

   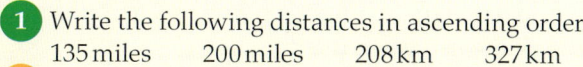

   Two students, Kwasi and Marita, are discussing what the function machine to convert miles to kilometres is.
   Kwasi writes down this inverse function machine.

   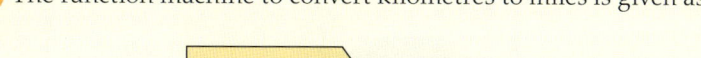

   Marita writes the following one:

   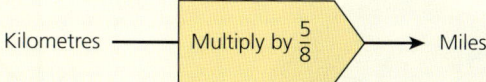

   a Comment on the accuracy of their function machines. Justify your answer.
   b The distance from the Earth to the moon is 240 250 miles. Convert the distance to kilometres.

3. A marathon is officially 42.195 km.
   A runner has run 18.1 miles. How far has she still got left to run?
   Give your answer
   a in kilometres
   b in miles.

4. The approximate multiplier to convert km to miles is to multiply the number of km by $\frac{5}{8}$. A more accurate conversion, correct to 5 significant figures, is to multiply the number of km by 0.621 37
   The distance between London and Paris for a cycle race is advertised as being 344 km.
   a What difference in miles do the two conversions give?
     Give your answer to 1 decimal place.
   b A cyclist says that he will use the 0.621 37 conversion as it will mean he has to cycle less. Is his logic correct? Justify your answer.

SECTION 3

Bearings

In the days when travelling and exploration were carried out on the world's oceans, compass **bearings** (directions), like those shown in the diagram, were used.

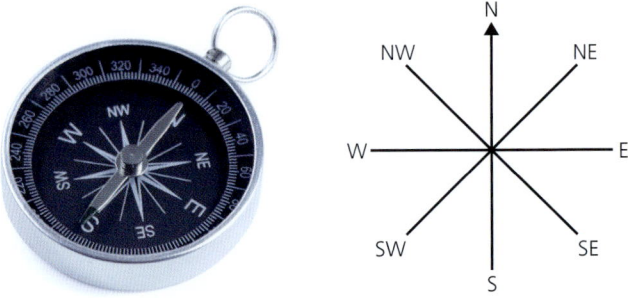

For greater accuracy, extra points were added midway between each of the existing eight points. Midway between north and north-east was north-north-east, midway between north-east and east was east-north-east, and so on. This gave the 16-point compass.

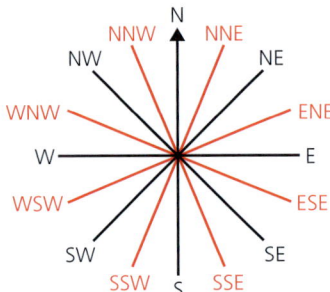

This was later extended to 32 points and even 64 points.

Today, however, this method of compass bearings has been replaced by a system of three-figure bearings. North is given a bearing of zero and 360° in a clockwise direction is one full rotation.

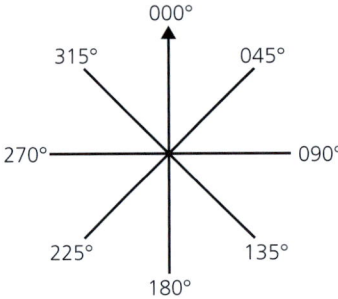

28 Distances and bearings

Measuring bearings

This diagram shows the positions of two boats, A and B.

To measure the bearing from A to B, follow these steps.
- As the bearing is being measured from A, draw a north arrow at A.

- Draw a straight line from A to B.

- Using a protractor or angle measurer, measure the angle from the north line at A to the line AB in a clockwise direction.

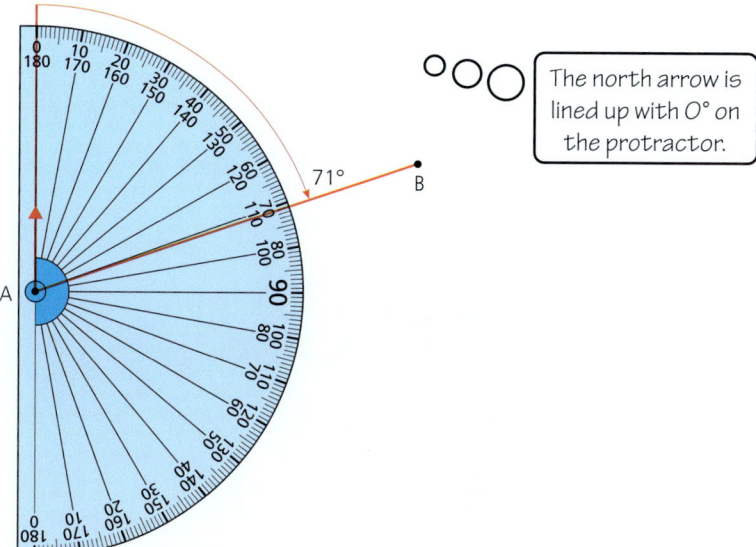

The north arrow is lined up with 0° on the protractor.

- The bearing is the angle written as a 3-digit number. Therefore, the bearing from A to B is 071°.

> **LET'S TALK**
> Why might it have been decided to always write bearings as a 3-digit number, even for angles less than 100°?

SECTION 3

Exercise 28.2

Trace each of the diagrams below. On your diagram, use a protractor to measure each of the following bearings.

1 Bearing from A to B

2 Bearing from X to Y

3 Bearing from P to Q

4 Bearing from M to N

5
 a Bearing from A to B
 b Bearing from C to B
 c Bearing from A to C

6
 a Bearing from E to D
 b Bearing from E to F
 c Bearing from G to F
 d Bearing from F to D

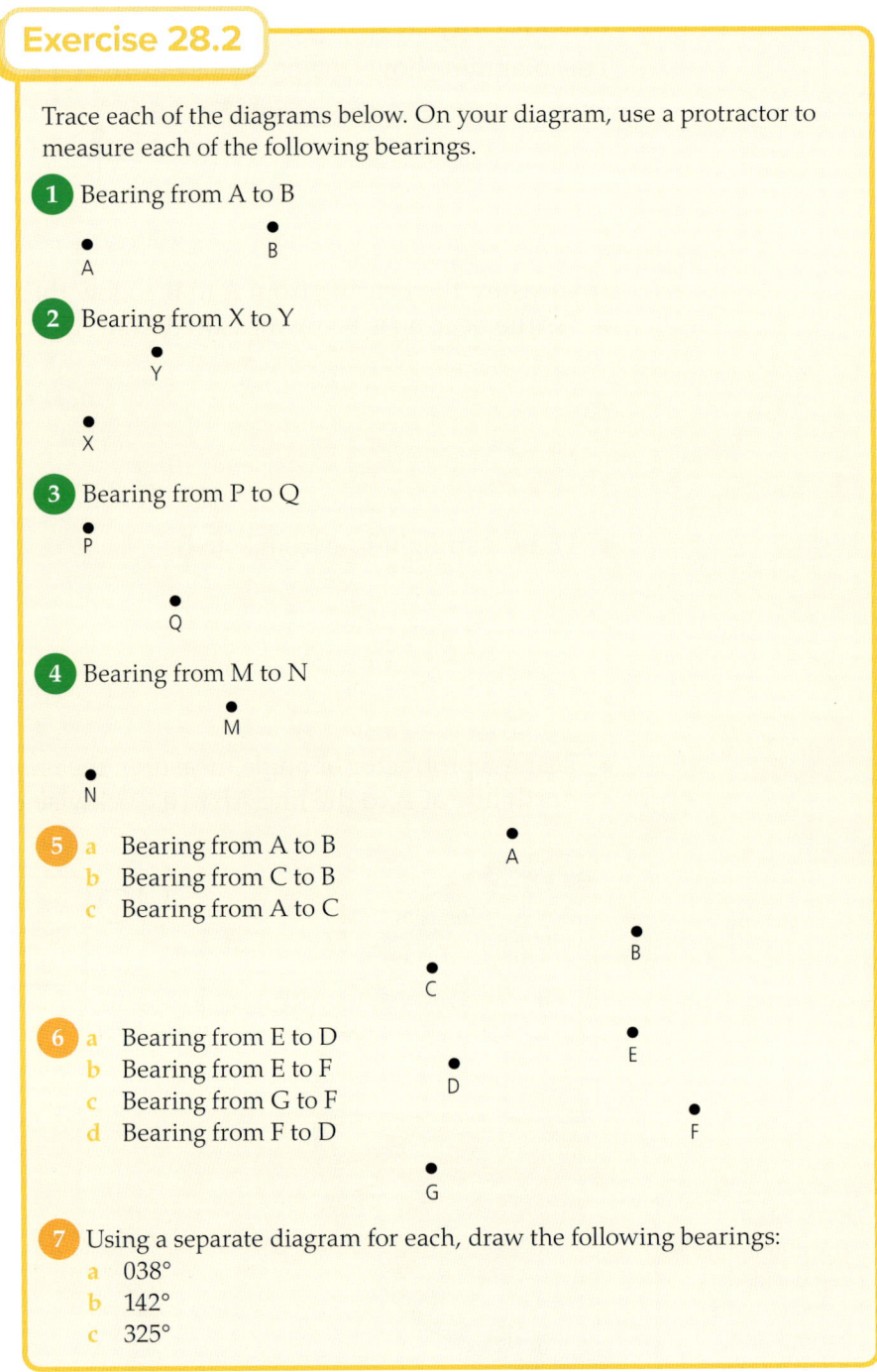

7 Using a separate diagram for each, draw the following bearings:
 a 038°
 b 142°
 c 325°

Now you have completed Unit 28, you may like to try the Unit 28 online knowledge test if you are using the Boost eBook.

29 Ratio

- Understand and use the relationship between ratio and direct proportion.
- Use knowledge of equivalence to simplify and compare ratios (different units).
- Understand how ratios are used to compare quantities to divide an amount into a given ratio with two or more parts.

LET'S TALK
Can you recall some of the uses of ratio in the real world?

Ratio and proportion

A ratio shows the relative sizes of two numbers. One number can be expressed as a multiple of the other number, as a part of the other number, as several parts of it or as a percentage of it.

Equivalent ratios

If two ratios are equivalent, then they are in proportion. For example, the ratios 3 to 6 and 10 to 20 are equivalent (in proportion) because 3 is half of 6 and 10 is half of 20.

> **Worked example**
>
> a Are the ratios 8 : 12 and 48 : 72 equivalent?
>
>   8 is two-thirds of 12 and 48 is two-thirds of 72, so the ratios are equivalent.
>
> b Are the ratios 18 to 12 and 60 to 40 in proportion?
>
>   18 is one and a half times 12 and 60 is one and a half times 40, so the ratios are in proportion.
>
> c Are the ratios 24 to 8 and 12 to 3 in proportion?
>
>   24 is 3 times 8 and 12 is 4 times 3, so the ratios are not in proportion.

SECTION 3

Exercise 29.1

1 Which of the following pairs of ratios are in proportion? Give reasons for your answers.
  a 1 to 5 and 4 to 20
  b 2 to 6 and 3 to 9
  c 5 to 25 and 15 to 75
  d 18 to 12 and 40 to 60
  e 7 to 21 and 21 to 42
  f 9 to 18 and 18 to 9

2 The square below has been shaded in three different colours.

  a Copy the rectangle below and shade it in the same proportion as the square.

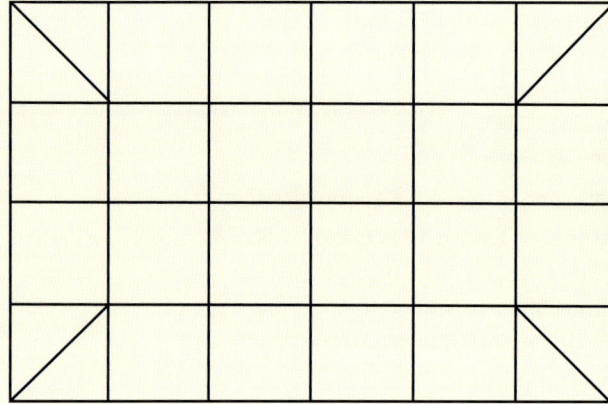

  b Justify your answer to part (a).

Simplifying ratios

Like fractions, ratios can be simplified. For example,

$4:8$ simplifies to $1:2$, as 4 is the **highest common factor** of both numbers and $4 \div 4 = 1$ while $8 \div 4 = 2$.

$14:21$ simplifies to $2:3$, as 7 is the highest common factor of both numbers and $14 \div 7 = 2$ while $21 \div 7 = 3$.

29 Ratio

Worked example

a Simplify the ratio 2 litres : 750 ml.
First write both quantities in the same units.
2 litres = 2000 ml
2000 ml : 750 ml simplifies to 8 : 3, since 250 is the highest common factor of both numbers and 2000 ÷ 250 = 8 while 750 ÷ 250 = 3.

It is important that the quantities are in the same units as 2000 : 750 is a very different ratio from 2 : 750.

LET'S TALK
The example writes both quantities in ml. What would the answer be if instead they were written in litres?

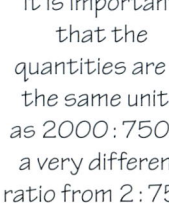

Exercise 29.2

Simplify the ratios in questions 1–10.

1. 27 : 36
2. 45 : 75
3. 28 : 70
4. 20 minutes : 1 hour
5. 3 kg : 750 g
6. 8 m : 75 cm
7. 7 litres to 750 ml
8. $2 to 20 cents
9. 3 hours to 1 day
10. 8 weeks to 1 year

11. A half-marathon is a distance of 21 km (to 2 s.f.).
A runner has entered a half-marathon race. After two hours, he has run a total of 18 500 m.
Write the ratio of distance left to run : distance already run in its simplest form.

12. A blue paint (A) is a mixture of blue : white in the ratio 25 : 60.
A second blue paint (B) is a mixture of blue : white in the ratio 40 : 88.
Which of the paints is the darker of the two? Justify your answer.

13. A tile pattern has three different coloured tiles as shown:
The designer wanted the ratio of blue : red : yellow in its simplest form to be 1 : 2 : 5.

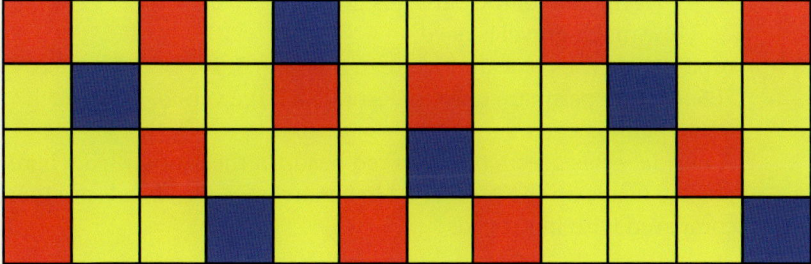

a Explain why the design is in the incorrect ratio.
b Which colours have the incorrect number of tiles? Justify your answer.

SECTION 3

When information is given as a ratio, the method of solving the problem is the same as when working with fractions.

> **Worked example**
>
> Copper and nickel are mixed in the ratio 7:4 to form a metal alloy.
> If 550 g of alloy is produced, how much copper is used?
>
> Nickel : Copper
>
>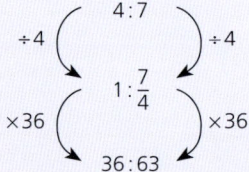
>
> A ratio of 7:4 means that the overall quantity has been split into 11 parts.
>
> 7 parts copper means that $\frac{7}{11}$ of the alloy is copper.
>
> Therefore the amount of copper is $\frac{7}{11} \times 550 = 350\,g$

> **Exercise 29.3**
>
> 1. The ratio of girls to boys in a class is 6:5. There are 18 girls.
>   a. What fraction of the class is boys?
>   b. How many boys are there?
>   c. If the ratio of girls to boys in the class had been 24:20, how would it have affected your answers to a and b above? Give a **convincing** argument for your answer.
> 2. Sand and gravel are mixed in the ratio 4:3 to make ballast. 140 kg of ballast is used on a building job.
>   a. What fraction of the ballast is gravel?
>   b. How much gravel is used?
> 3. A paint mix uses blue and white in the ratio 3:10. 286 litres of paint are used to decorate a large venue. How much white paint is used?
> 4. A necklace has green, blue and red beads in the ratio 2:3:5. There are 240 beads on the necklace. How many more red beads are there compared with blue beads?

Dividing a quantity in a given ratio

We can also use the unitary method to divide a quantity in a given ratio.

29 Ratio

> **Worked example**
>
> A piece of wood is 150 cm long. It is divided into two pieces in the ratio 7:3.
>
> How long is each piece?
>
> A ratio of 7:3 means that you need to consider the wood as 10 parts.
>
> One piece of wood is made of 7 parts; the other piece is made of 3 parts.
>
> 10 parts are 150 cm long.
>
> So, 1 part is 150 cm ÷ 10 = 15 cm long.
>
> 7 parts are 7 × 15 cm = 105 cm long and 3 parts are 3 × 15 cm = 45 cm long.

Exercise 29.4

1.
   a Divide 250 in the ratio 3:2.
   b Divide 250 in the ratio 6:4
   c Compare your answers to parts (a) and (b) above. Justify any similarity between the two. Find another ratio that produces the same result.

2.
   a Divide 144 in the ratio 1:2.
   b Divide 144 in the ratio 12:24.
   c Compare your answers to parts (a) and (b) above. Justify any similarity between the two. Find another ratio that produces the same result.

3. Divide 10 kg in the ratio 2:3.

4. Divide 8 m in the ratio 3:13.

5. A paper strip is coloured as follows:

   State whether the following statements are true or false. Justify your answers.
   a The ratio of orange:blue squares is 6:4.
   b The proportion of blue squares is $\frac{4}{6}$.
   c The ratio of blue:orange squares is 2:3.
   d The proportion of orange squares is $\frac{6}{10}$.
   e The proportion of blue squares is $\frac{2}{5}$.

SECTION 3

6. Divide 1 hour in the ratio 10:14. Give your answer
   a in minutes
   b in its simplest form.
7. Divide 2 kg in the ratio 9:21. Give your answer
   a in grams
   b in its simplest form.
8. Divide 1 m in the ratio 10:15:25. Give your answer
   a in cm
   b in its simplest form.
9. Divide 1 hour in the ratio 5:6:9. Give your answer
   a in minutes
   b in its simplest form.
10. Divide 2 km in the ratio 6:16:18. Give your answer
    a in metres
    b in its simplest form.
11. The size of angles in a triangle ABC are in the ratio 2:3:5 respectively.
    a Calculate the size of angle C.
    b Justify your answer.
12. The two rectangles, P and Q, below have dimensions as shown:

The ratio of the area of P:Q is 8:5. Calculate:
a the value of x
b the ratio of their perimeters, giving your answer in its simplest form.
c Find another pair of rectangles, A and B, whose areas are also in the ratio 8:5.
d Harry **conjectures** that the ratio of the perimeters of P and Q will definitely be the same as the ratio of the perimeters A and B. Is Harry correct? Give a **convincing** reason for your answer.

LET'S TALK

Is pay always in direct proportion to the number of hours worked? Think of examples that support your ideas.

Direct proportion

Workers building houses may be paid for the number of hours they work. Their pay is in **direct proportion** to the hours they work – more hours, more pay.

Workers laying bricks may be paid for the number of bricks they lay, not for the time they work. Their pay is in direct proportion to the number of bricks they lay.

29 Ratio

Worked example

A machine for making cups makes 500 cups in 20 minutes.

How many cups does it make in 3 hours?

In 20 minutes, 500 cups are made.

So, in 1 minute 500 ÷ 20 = 25 cups are made.

3 hours is 180 minutes.

In 3 hours 25 × 180 = 4500 cups are made.

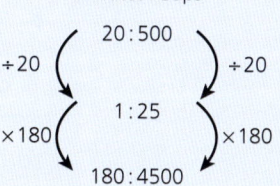

The method used in the example above is called the unitary method. In it, you work out the ratio in the form $1:n$ and then you multiply by the appropriate value.

Exercise 29.5

Use the unitary method to solve these problems.

1. A combine harvester produces 9 tonnes of grain in 6 hours. How many tonnes does it produce in 54 hours?
2. A heater uses 3 units of electricity in 40 minutes. How many units does it use in 2 hours?
3. A machine prints 1500 newspapers in 45 minutes. How many does it print in 12 hours?
4. A bricklayer lays 1200 bricks in an average 8-hour day. How many bricks does he lay in a 40-hour week?
5. A machine puts tar on a road at the rate of 4 metres in 5 minutes.
   a. How long does it take to cover 1 km of road?
   b. How many metres of road does it cover in 8 hours?

Now you have completed Unit 29, you may like to try the Unit 29 online knowledge test if you are using the Boost eBook.

Reading and interpreting graphs

- Read and interpret graphs with more than one component. Explain why they have a specific shape and the significance of intersections of the graphs.

You will already be familiar with how to interpret certain types of graphs such as frequency diagrams, pie charts and scatter graphs. This unit will look specifically at graphs that can be split into different sections and looks at the significance of their different shapes.

Motion graphs

Describing the motion of an object is often best visualised using a distance–time graph. This shows how far the object has moved over time.

Worked example

Look at the graph below. It shows the distance travelled in a straight line (in metres) by two objects, A and B, over a period of 10 seconds.

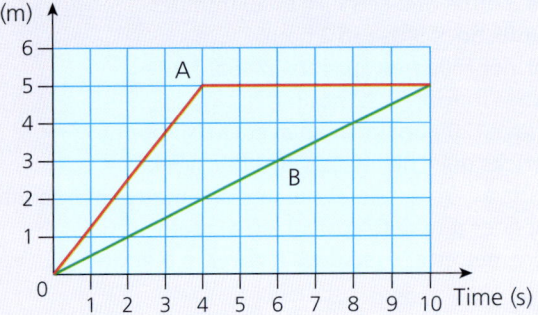

LET'S TALK
What do the different sections of the graphs describe about the object's movements?

a How far did object A travel in the first four seconds?
  In the first four seconds object A travelled 5 m.

b Describe the motion of A between the 4th and 10th second.
  As the graph is horizontal it means object A has not travelled any distance, therefore it is stationary.

c Which object is travelling more quickly in the first four seconds? Justify your answer.
  Object A, because it has travelled 5 m in four seconds, while object B has only travelled 2 m in four seconds.

LET'S TALK
What does a straight diagonal line say about an object's movement?

30 Reading and interpreting graphs

Exercise 30.1

1 Two students, X and Y, run in a straight line towards a ball 10 m away. Both start at the same time. A graph of their motion is shown below:

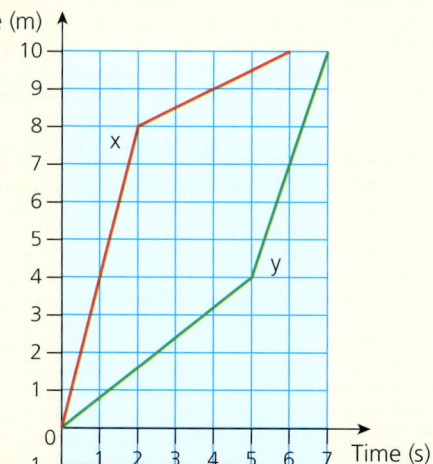

LET'S TALK

What does a steeper line imply about an object's motion?

a Which player reaches the ball first? Justify your answer.
b Which player runs the fastest in the first two seconds? Justify your answer.
c How far away from the ball was the slowest student when the fastest student reached it? Explain how this can be seen from the graph.

2 Two trains, T1 and T2, leave the same station and head towards the same final destination 100 km away. One of the trains is an express train, the other a slower train. The graph below shows their journeys.

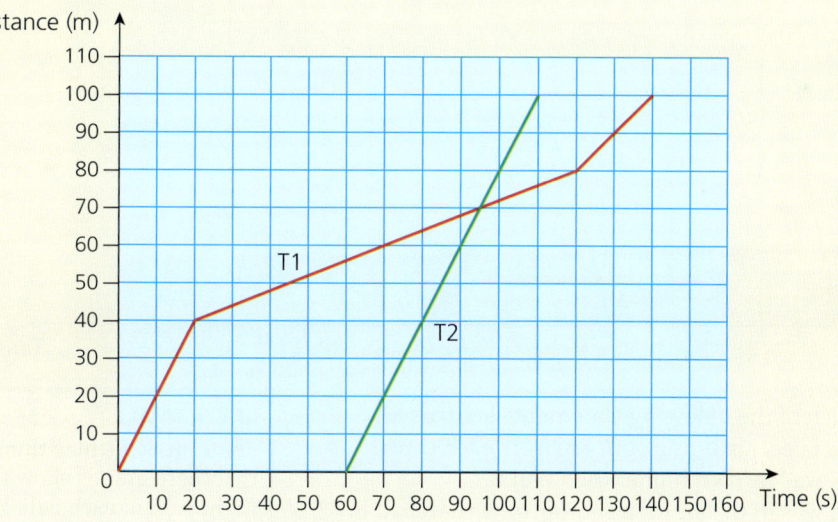

257

SECTION 3

   a Which train is the express train? Justify your answer.
   b Did both trains set off at the same time? Justify your answer.
   c If train T2 departed at 8:00 a.m., what time did it arrive at its destination?
   d What time did T1 arrive at the same destination?
   e i) Approximately at what time did T2 overtake T1? Justify your answer.
       ii) How far from the start did this happen?
   f Train T1's journey can be split into three stages. At what stage was it travelling the fastest? Justify your answer.

3 During an experiment, a ball (A) is thrown vertically upwards into the air. Shortly afterwards, a second ball (B) is also thrown vertically up into the air. The graph below shows their height above the ground over time.

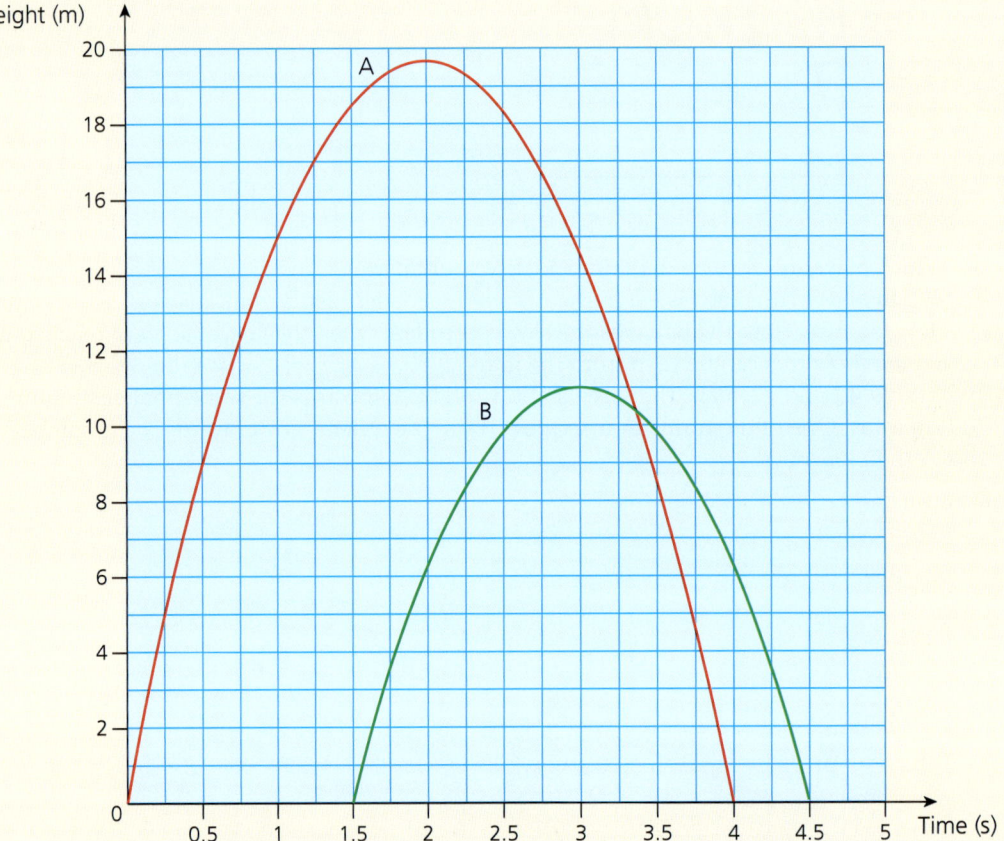

   a State which of the following statements are true and which are false, justifying your answers each time.
      i) Ball A was thrown higher than ball B.
      ii) Ball A was in the air for a shorter time than ball B.
      iii) Ball B was thrown into the air 1.5 seconds after ball A.

LET'S TALK
Some students may think that these graphs show the flight path of each ball. Why might they think that and why is this wrong?

iv) The balls are at the same height at one point during the experiment.
v) Ball A reaches its highest point 2 seconds after it is thrown into the air.
vi) Ball B reaches its highest point 3 seconds after it is thrown into the air.

b i) How long after ball A is thrown up into the air is it at the same height as ball B? How can you tell that from the graph?
ii) At what height above the ground do the balls pass each other?
iii) Do the balls pass each other on the way up or on the way down? Justify your answer.

Travel graphs are not the only types of scenario which can be represented by graphs involving more than one component. The following exercise looks at some other possible cases.

Exercise 30.2

1. Two identical water barrels, P and Q, are shown on the graph below. The tap near the bottom of barrel P is left open and the water pours out at a constant rate, while water is being added to Q from a tap above it.

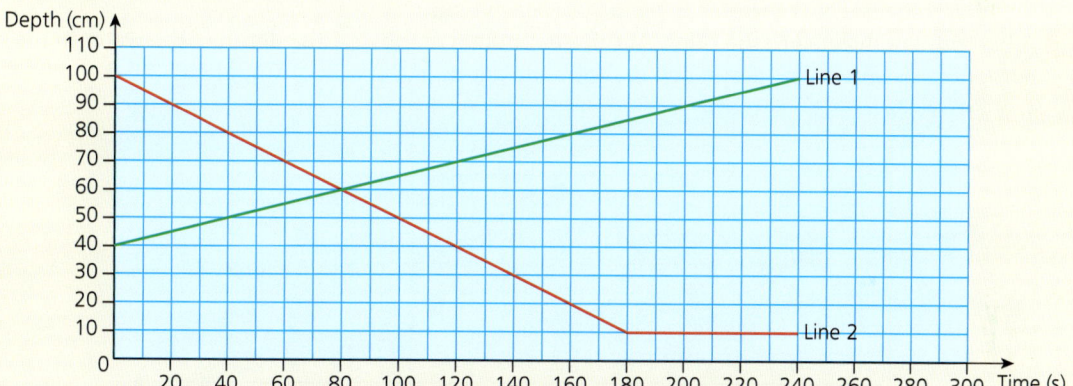

The graph shows the depth of water in both barrels over time:
a Which graph belongs to barrel P and which to barrel Q? Justify your choice.
b How much water was in barrel Q at the start?
c What is the approximate difference in water levels between the two barrels 3 minutes after the start?
d Is there a time when both barrels have the same depth of water in them? If so, when?

SECTION 3

2 A cup of hot coffee is placed in a cup on a table. In the same room, an iced drink is poured into a glass and also placed on the table.

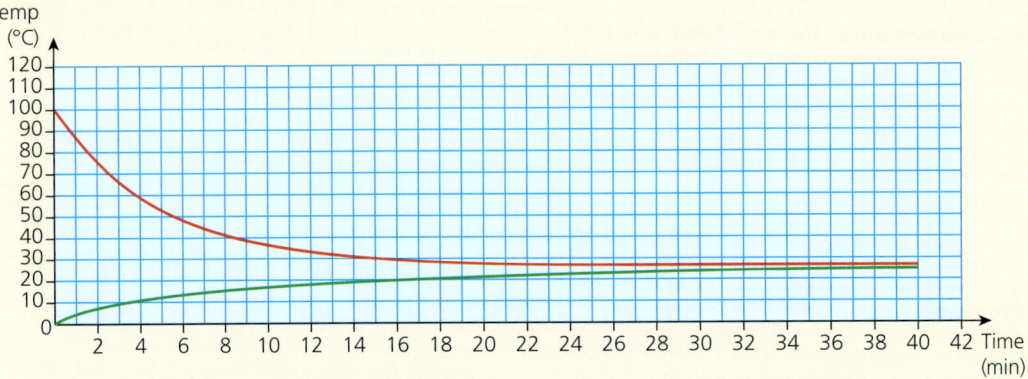

Their temperature is recorded over time.
A graph of their change in temperature is shown:
  a Which of the lines above represents the cup of coffee? Justify your answer.
  b Does the coffee cool at the same rate over the 40 minutes? Justify your answer.
  c Two students are discussing the graphs. One of them says that it is possible to deduce the room temperature from the graphs. His friend says that it is not possible to deduce this. Which statement is correct? Explain your reasoning.

3 A harbour is subjected to the tides. At different points of the day, the depth of the water in the harbour is recorded. The graph below shows the depth of the water throughout one day.

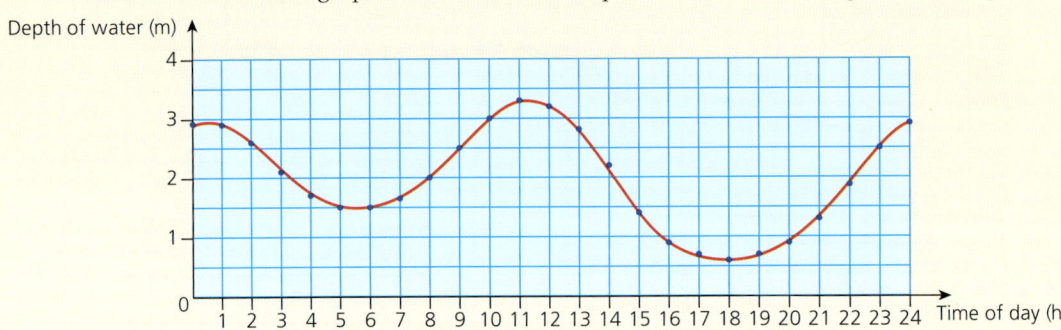

  a At what time of the day is the water level the lowest?
  b What was the approximate difference in height of the water between the highest and lowest points?
  c A boat can only sail into the harbour if the water level is above 2 m. Approximately, at what times of this particular day can the boat not sail into harbour?
  d A large boat can only sail into the harbour if the water is above 3 m.
     A large boat carrying passengers has to sail into the harbour, dock, unload its passengers, pick up new passengers and sail out again. What's the maximum amount of time it has to do this on this particular day? Explain your reasoning.

Now you have completed Unit 30, you may like to try the Unit 30 online knowledge test if you are using the Boost eBook.

Section 3 – Review

1. Look at the pattern of square tiles below:

   a Draw the next pattern in the sequence.
   b Write down the term-to-term rule for the number of white squares.
   c Write down the rule for the nth term of the number of white squares.
   d Use your rule for the nth term to calculate the number of white squares in the 20th pattern of the sequence.

2. During the day, the maximum temperature recorded at a ski resort is 6 °C. At night, the temperature drops by 450%.
   a Calculate the temperature at night.
   b What is the absolute change in temperature between daytime and night-time?

3. Two identical cylindrical grain stores are joined and have dimensions as shown opposite.
   Using a scale of 1:200, draw front, side and plan elevations of the grain stores.

   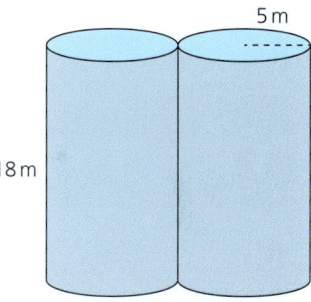

4. Write each of the following machine functions algebraically, using x and y as the input and output respectively.

   a In → Add 2 → Divide by 5 → Out

   b In → Divide by 4 → Subtract 3 → Out

SECTION 3

5 The axes below show a quadrilateral ABCD:

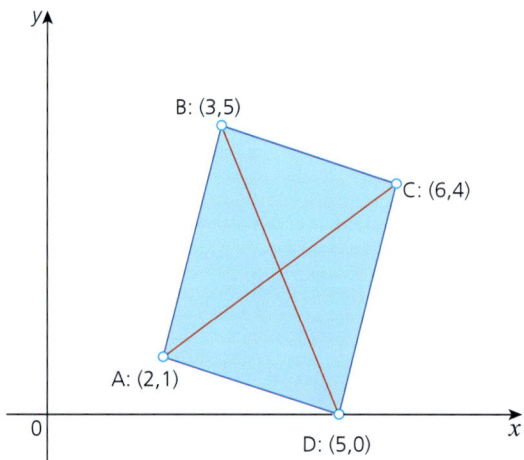

Prove that ABCD is a parallelogram.

6 Without a calculator, calculate the value of $\sqrt{0.49}$. Show your method clearly.

7 A straight line has equation $y = \frac{1}{3}x - 4$. The coordinates of five points are given below.

(9,−1)   (−12,−8)   (3, 3) $\left(0, -3\frac{2}{3}\right)$ (−15,1)

Which of the five points lie on the line?

8 The bearing of A to B is given as 058°. The distance AB is 6 cm.
  a Draw the diagram showing the position of B relative to A.
  b Use your diagram to work out the bearing of B to A.

9 Copy the 8×8 grid below.

Colour the grid using three colours in the ratio 1:2:5.

10 A gas company charges customers by the number of cubic metres of gas used. It has two tariffs to offer customers.
Tariff A is $20 per month fixed charge and then $1.50 for each cubic metre of gas used.
Tariff B is $40 per month fixed charge and then $0.80 for each cubic metre of gas used.

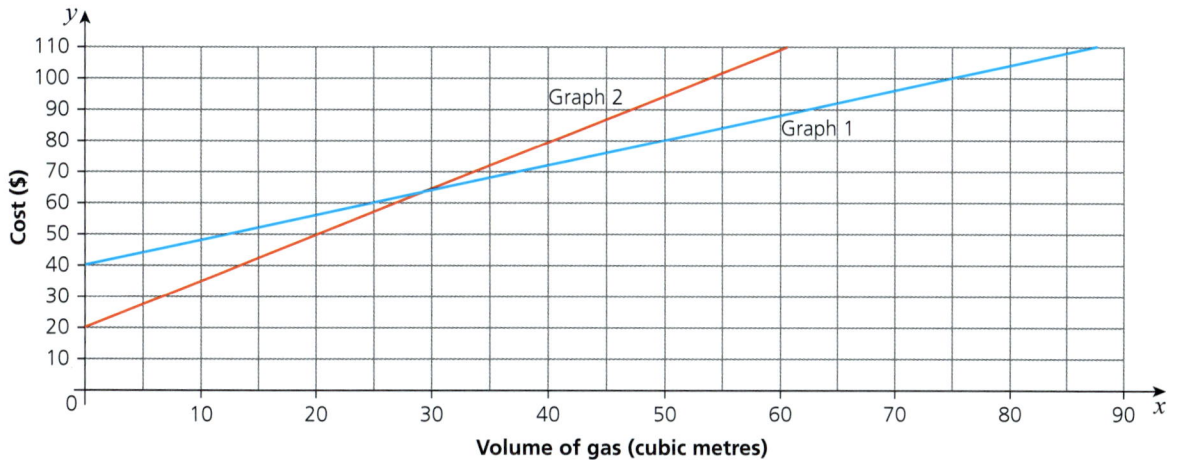

a Which of the graphs is for tariff A and which for tariff B? Justify your choices.
b A family uses on average 60 m³ of gas per month. Which tariff should they choose?
Justify your answer.
c Another family uses on average 30 m³ of gas per month. Which tariff should they choose? Justify your answer.

Glossary

Absolute change The actual difference between two values over a period of time.

Adjacent Next to each other. E.g. adjacent sides are sides that are next to each other.

Alternate angles When two parallel lines are intersected by a straight line, the two opposite angles formed are known as alternate angles and are equal in size. They can be found in the Z-shape formation of the lines.

Angle A measure of an amount of turn. It is measured in degrees.

Base (number) A number that is raised to a power. E.g. in 5^3, the 5 is the base number.

Bearing An angle measured clockwise from North. It is written as a three-digit number.

Bisect/bisector To divide something in half, e.g. bisect an angle or bisect a line.

Categorical data Data that are used as labels rather than quantities, such as a favourite colour.

Centre of enlargement Enlarging an object requires a centre of enlargement. The distance from the centre of enlargement to each point on the object is multiplied by the scale factor to produce the enlarged shape.

Centre of rotation A point about which an object with rotational symmetry is turned.

Circumcircle A circle which passes through all the vertices of a polygon.

Coefficient The number with which we multiply the **variable** in question.

Column vector A column vector describes a translation. E.g. $\begin{pmatrix} 2 \\ -5 \end{pmatrix}$ implies an object has moved 2 units to the right and 5 units down.

Commutative Commutative multiplication means that the multiplication can be done in any order and the answer will be the same, e.g. $2 \times 3 \times 4 = 4 \times 2 \times 3 = 3 \times 4 \times 2$ etc.

Congruent Two shapes are congruent if they are exactly the same size and shape.

Constant A quantity whose value stays the same, as opposed to a variable whose value is **variable**.

Constant cross-sectional area If a 3D object is the same all the way through its body, it is said to have a constant cross-sectional area.

Continuous data Data that can take any values. Examples include time, height and mass. Because continuous data can take any value, there are an infinite number of possible values.

Convex polyhedron (polyhedra) A polyhedron is a 3D shape where all its faces are polygons. A convex polyhedron is one in which if two corners of the polyhedron are joined by a straight line, the line always passes through the inside of the shape.

Corresponding angles When two parallel lines are intersected by a straight line, the two angles found in the same relative position are corresponding angles and are equal in size. They can be found in the F-formation of the lines.

Cross-section The shape formed when a 3D object is sliced through.

Cube number A number that is made by multiplying an integer by itself three times. For example, $5 \times 5 \times 5 = 125$. Therefore, 125 is a cube number.

Cubing Multiplying a number by itself three times.

Glossary

Denominator The bottom number in a fraction.

Direct proportion Two quantities are in direct proportion if their ratio is constant.

Discrete data Numerical data that cannot be shown in decimals, for example, the number of children in a classroom.

Distributive law This refers to a law of multiplication. Multiplying a group of numbers being added together gives the same answer as multiplying each of the numbers first and then adding them together,
e.g. $5(3 + 8) = (5 \times 3) + (5 \times 8)$.

Dodecahedron A convex polyhedron with 12 flat faces.

Enlargement A type of transformation. All lengths of the original shape are multiplied by the same value known as the scale factor of enlargement.

Equally likely Events are equally likely if they each have the same chance of happening.

Equation A mathematical statement stating that two quantities are the same.

Equation of a straight line An equation which describes the relationship between the x and y coordinates of all the points on the line. E.g. on a line with equation $y - 2x$ the y-coordinates of every point are twice the value of the x-coordinates.

Equidistant The same distance away from a point.

Euler's formula This is a formula, proved by the mathematician Leonhard Euler, describing the relationship between the number of vertices (V), faces (F) and edges (E) of any polyhedron. It states that $V + F - E = 2$.

Event Used in probability, it refers to something that has happened.

Exact form Giving an answer in exact form means to leave irrational numbers such as π in the answer rather than giving it as a decimal approximation, e.g. 3π rather than 9.42 (2 d.p.).

Expansion Expanding an expression involving brackets means to multiply out the brackets, e.g. $2(x + 3)$ when expanded is $2x + 6$, as the 2 has multiplied both terms inside the bracket.

Expression A mathematical statement written using algebra. It does not include an equals sign. For example, $3(x + y)$.

Exterior angle If an edge of a polygon is extended, the angle formed between the extended line and the next edge is an exterior angle.

Face A single flat surface on a three-dimensional shape.

Factor A number that goes into another number exactly. For example, 5 is a factor of 20 because $20 \div 5 = 4$.

Factorisation The process of removing the highest common factor from some terms. E.g. factorising $3x + 12 = 3(x + 4)$ as 3 is the highest common factor of $3x$ and 12.

General form The general form of a straight line shows the way any straight line can be written, e.g. any straight line can be written in the form $y = mx + c$.

Gradient The slope or steepness of a line.

Hierarchy This describes the position of numbers or shapes relative to each other in terms of their properties.

Highest common factor (HCF) The largest number that is a factor to two (or more) other numbers.

Hypotenuse The longest side of a right-angled triangle. It is the side opposite the right angle.

Glossary

Image The name of a shape after its position has been transformed (changed) by rotation, reflection or enlargement.

Improper fraction A fraction where the numerator is bigger than the denominator.

Independent events Outcomes that do not affect each other. E.g. when rolling two dice, the result of one will not affect the result of the other, so they are independent events.

Integer A whole number (not a fraction) that can be positive, negative or zero. Therefore, the numbers 10, 0, −25 and 5148 are all integers.

Intercept Where a line crosses an axis on a graph. Due to the equation of a straight line, the intercept is often used just to mean where it crosses the y–axis.

Intersect Where two lines cross.

Inverse function The inverse function is the function that undoes the effect of the original function.

Line A line is a straight one-dimensional figure with no thickness and with an infinite length. However, most of the time when referring to a line, people are actually meaning a line segment.

Line segment A straight line between two points.

Lowest common multiple (LCM) The lowest common multiple of two numbers is the smallest whole number which both go into.

Map In a transformation, the mapping shows how an image is transformed (either by reflection, rotation, translation or enlargement) to its new shape and position.

Mathematical operations These are mathematical processes applied to numbers. The four most common ones are addition, subtraction, multiplication and division.

Mean A type of average. Add all the values together and divide by how many numbers there are.

Median A type of average. It is calculated by writing all the numbers in order and then selecting the middle number.

Midpoint Halfway point on a straight line.

Mirror line The line of reflection symmetry in a two-dimensional shape or pattern.

Mixed number A number which contains an integer and a proper fraction, e.g. $2\frac{3}{5}$

Modal group In grouped data, this is the group with the largest frequency.

Mode A type of average. It is the number that occurs the most often. Also known as the modal value.

Model (modelling) Modelling a situation means to represent it, often in a simpler way, in order to study the likely behaviour of the actual event.

Mutually exclusive events Events that cannot happen at the same time, e.g. getting a head and a tail at the same time when flipping a coin.

nth term A formula for the nth term of a sequence enables you to find any term in that sequence.

Natural number A whole number bigger than zero. E.g. 1, 2, 3, 4 etc…

Negative integer A whole number less than zero, e.g. −5.

Net In geometry, a net is the 2D shape that can be folded to make a 3D object.

Numerator The top number in a fraction.

Object An object is moved from its original position to a new position in a transformation.

Glossary

Order of rotational symmetry The number of times a shape or pattern will look the same when rotated a full 360°.

Parallel Always the same distance apart and never touching.

Pentagonal-based pyramid A pyramid where the base is in the shape of a pentagon.

Percentage A number or ratio given out of 100.

Perpendicular Perpendicular lines are at right angles to each other.

Pie chart scale A circular scale given in either degrees or percentages, used to draw pie charts.

Polygon A flat two-dimensional shape with straight sides. Quadrilaterals, triangles and hexagons are all examples of polygons.

Polyhedra/polyhedron A three-dimensional shape in which all its faces are polygons.

Population The set of members under study where samples are taken.

Positive integer A whole number greater than zero.

Possible outcomes All the possible results when doing an experiment, e.g. when rolling an ordinary dice, the possible outcomes are 1, 2, 3, 4, 5 or 6.

Prime factor A factor of a number which is also a prime number. E.g. the prime factors of 20 are 2 and 5 as both go into 20 and are also prime numbers.

Prime number A number which has only two factors, 1 and the number itself.

Prism A 3D shape that has a constant cross-section.

Proper fraction A fraction where the numerator is less than the denominator.

Proof A mathematical argument that shows a statement is logically true.

Protractor (angle measurer) Mathematical equipment used to measure the size of an angle. A protractor measures up to 180° while an angle measurer measures up to a full 360°.

Quadrant A quarter of a circle and is the shape formed between two radii at right angles to each other and the arc (part of the circumference). It is a specific type of sector.

Quantitative data Data representing an amount that can be measured, such as mass, height, number of people etc.

Ratio How two or more quantities are related to each other.

Rational number Any number that can be written as a fraction. These include whole numbers and terminating and recurring decimals.

Rearranging Changing the order of a formula to make a different variable the subject.

Recurring decimal A decimal number which repeats itself.

Reflection A type of symmetry where one side of a mirror line is exactly the same as the other.

Regular pentagon A five-sided polygon where all the sides are of equal length and all the angles of equal size.

Representative sample A selection chosen from the population, which has similar properties to the whole population.

Rotational symmetry When an *object* is rotated around a centre point (turned) a number of degrees, the object appears the same. The order of symmetry is the number of positions the object looks the same in a 360° rotation.

Sample A group chosen from a *population*.

Sample size The size of the group chosen from a *population*.

Sample space diagram A table which shows all the possible outcomes of an event.

Glossary

Scale factor of enlargement How much each side of the original shape has been multiplied by to produce the enlarged shape.

Sector A fraction/part of a circle. It is the shape formed between two radii and the arc. Note that a quadrant is a specific type of sector.

Segment (line) Part of a straight line that is between two points.

Sequence A group of numbers or shapes that follow a rule or pattern, e.g. 2, 4, 6, 8, 10 is a sequence of numbers that increases by 2.

Significant figures The number of significant/meaningful digits in a number. E.g. 4.23 has 3 significant figures while 0.0023 only has 2 significant figures.

Similar Shapes that are mathematically similar are the same shape but a different size from each other. Their angles will be the same as each other and the lengths of their sides will stay in proportion.

Solve To determine the solutions to a problem.

Square number The result of multiplying a whole number by itself. For example, 5 × 5 = 25, so 25 is a square number.

Squaring Multiplying a number by itself.

Stellated Star shaped.

Stem-and-leaf diagram A type of statistical diagram for numerical data. Usually, the last digit of the numbers form the leaf part of the diagram while the rest of the digits form the stem part.

Supplementary angles Two angles that add up to 180°.

Translation Translation is a sliding movement. To describe a translation, you need to give how far the shape has moved horizontally and how far it has moved vertically.

Term Numbers in a sequence.

Terminating decimal A decimal which has an end, e.g. 4.2.

Term-to-term rule A rule that describes how to get from one term to the next in a sequence.

Theoretical probability The likelihood of something happening in theory. For example, the theoretical probability of getting a tail when flipping a coin is $\frac{1}{2}$.

Time-series graph A line graph in which the x-axis is time.

Tree diagram A diagram used in probability calculations. The branches of the tree show all the possible outcomes and their probability.

Triangle number A number that belongs to a special sequence called the sequence of triangle numbers. The first six terms of this sequence are 1, 3, 6, 10, 15, 21.

Triangular prism A 3D shape that has a constant cross-section in the shape of a triangle.

Two-way tables A table that shows the frequency of two types of data.

Variable A mathematical quantity whose value can vary.

Vector A quantity that has both a direction and a size.

Vertically opposite angles Angles formed when two straight lines intersect, forming an 'X' shape. They are opposite each other and have the same size.

Zero index When a number has been raised to the power of 0. Any number raised to the power of 0 is equal to 1.

Index

2D shapes *see* two-dimensional shapes,
3D shapes *see* three-dimensional shapes

A

addition
 equations 43–4, 184–7
 mixed numbers 133–6
 order of operations 40
algebraic expressions 41–3
 deriving formulae 177–80
 expanding brackets 141–3
 expansion and factorisation 144–7
 and formulae 175–6
 history of mathematics 195
 language of algebra 141
alternate angles 167–9
angles
 alternate and corresponding angles 167–9
 bisecting 164–6
 of quadrilaterals 173–4
 of triangles 153–4, 170–2
averages 105–10

B

bearings 246–8
BIDMAS 36–40, 137
bisectors 162–6
brackets
 expanding 141–3
 order of operations 36–40

C

categorical data 48
centre of enlargement 125
centre of rotation 116
charts *see* graphs
Chinese history of mathematics 1
circles 31–5
circumference 31–5
coefficients 41
column vectors 224
combined events 148–52
commutative nature of multiplication 142
complementary events 90–2
congruence (shapes) 121, 125
constants 41
continuous data 61
convex polyhedra 75–6
coordinate pairs 234–7
corresponding angles 167–9
cross-sectional area 77
cube numbers 231–2
cube roots 232–3

D

data collection
 questionnaires 23–4
 sampling 19–23
data comparison/interpretation
 averages and the range 105–10
 interpreting further data 111–14
data representation 48
 decisions and interpretations 68–72
 frequency tables 49–50
 grouped frequency diagrams 51–4
 pie charts 54–60
 scatter graphs 63–8
 stem-and-leaf diagrams 61–3
 Venn diagrams 48–9
 see also graphs
data sampling 19–23
decimals
 place value 96–104
 terminating and recurring 88, 129–32
denominators 132
direct proportion 254–5
discrete data 48, 61
distance-time graphs 256–60
distances 244–5
distributive law 141
division
 calculations and worked examples 2–9
 indices 11–12
 mixed numbers 136–40
 numbers between 0 and 1 99–101, 102–4
 order of operations 40
 the zero index 13–14
dodecahedrons 76

E

elevation (shapes) 209–13
enlargement (shapes) 125–8

269

Index

equation of a straight line 238–9
equations 43–5, 184–9
  see also algebraic expressions
equidistance 164
Euler's formula 73–5
expansion (algebraic expressions) 144–7
expressions 41–3
exterior angles 171

F

factor trees 83–7
factorisation 144–7
factors 83–7
formulae 45–7
  deriving 177–80
  see also algebraic expressions
fractions
  mixed numbers and improper fractions 132–6
  multiplication and division involving mixed numbers 136–40
frequency tables 49–50
function machines 214–20
functions 214–20

G

general equation of a straight line 240–3
geometry see three-dimensional shapes; two-dimensional shapes
graphs
  coordinate pairs 234–7
  decisions and interpretations 68–72
  frequency tables 49–50
  general equation of a straight line 240–3
  grouped frequency diagrams 51–4
  linear graphs 237–9
  motion 256–60
  pie charts 54–60
  reading and interpreting 256–60
  scatter graphs 63–8
  stem-and-leaf diagrams 61–3
  Venn diagrams 48–9
grouped frequency diagrams 51–4

H

hierarchies
  number groups 87–9
  of quadrilaterals 15–18
highest common factors (HCFs) 84–5, 250
history of mathematics
  Chinese 1
  Indian 95
  Persian 195
the hypotenuse 154, 159
hypothesis 182

I

improper fractions 132–6
independent events 148–52
Indian history of mathematics 95
Indices 9–10
  laws of 10–12
  and roots 38–40
  the zero index 13–14
inequalities
  combined 191
  signs 189–90
integers 87
intercept (equation of a straight line) 241–3
intersection
  constructing triangles 153–62
  equation of a straight line 241–3
isosceles triangles 170

L

language of algebra 141
  see also algebraic expressions
line segments 221–3
linear graphs 237–9
lines
  coordinate pairs 234–7
  parallel and intersecting 167
  perpendicular bisectors and the midpoint 162–4
lowest common multiples (LCMs) 85

M

mathematical operations
  expressions 41
  order of operations 36–40
  see also division; multiplication
mean 105–10
median 105–10
midpoints, of line segments 221–3
mixed numbers
  improper fractions 132–6
  multiplication and division 136–40
modal group 105
mode 105–10
multiples 83
multiplication

Index

calculations and worked examples 2–9
  decimals 131–2
  distributive law 141
  expanding brackets 141–3
  indices 9–12
  mixed numbers 136–40
  numbers between 0 and 1 99–102
  order of operations 40
  the zero index 13–14
multiplication grids 2–3, 6
mutually exclusive events 148–52

N
natural numbers 87
negative integers 87
nth term 196, 200–3
number groups 87–9
numerators 132

O
operations *see* mathematical operations
order of operations 36–40

P
parallel lines 167
parallelograms 25–31
pentagonal-based pyramids 76
pentagons 76
percentage decreases 204–8
percentage increases 204–8
perpendicular lines 162–4
Persian history of mathematics 195
pie charts 54–60
place value 96–104
polygons 73

polyhedral 73–6
positive integers 87
prime numbers 83–4
prisms 77–9
probability
  complementary events 90–2
  experiments and simulations 181–3
  mutually exclusive and independent events 148–52
proper fractions 132
proportion
  direct 254–5
  ratio 249–54

Q
quadrants 35
quadrilaterals
  angles 173–4
  hierarchy of 15–18

R
range 105–10
ratio 249
  dividing a quantity in a given ratio 252–4
  equivalent ratios 249–50
  simplifying 250–2
rational numbers 87
recording data 48
recurring decimals 88, 129–32
reflection 115–21
representing data *see* data representation
right-angled triangles 154, 159–62
roots
  cube 232–3
  and indices 38–40
  square 229–31

rotation 121–3
rotational symmetry 116, 121–3

S
sample space diagrams 150
sampling *see* data sampling
scale factor of enlargement 125
scale (shapes) 209–13
scatter graphs 63–8
semicircles 33
sequences
  nth term 196, 200–3
  term-to-term rules 196–9
shapes *see* three-dimensional shapes; two-dimensional shapes
significant figures (s.f.) 96–8
square numbers 229
square roots 229–31
statistical calculations 108–10
stem-and-leaf diagrams 61–3
subtraction
  decimals 132
  equations 43–4, 184–7
  mixed numbers 133–6
subtraction (order of operations) 40
supplementary angles 167
surface area 80–2

T
term-to-term rules 196–9
terminating decimals 88, 129–32
terms 41
theoretical probability 181–3
three-dimensional shapes 73
  2D representations of 3D shapes 209–13
  convex polyhedra 75–6

271

Index

Euler's formula 73–5
surface area 80–2
triangular prisms 77–9
time series graphs 67–8
transformation of 2D shapes
   additional transformations 123–8
   reflection and rotation 115–21
   rotation about a point 121–3
translations 223–8
trapezia 25–31
travel graphs 256–60
tree diagrams 150
triangle numbers 196

triangles
   and angles 170–2
   constructing 153–62
   the hypotenuse 154, 159
   right-angled 159–62
triangular prisms 77–9
two-dimensional shapes
   2D representations of 3D shapes 209–13
   angles of a quadrilateral 173–4
   circles 31–5
   hierarchies of quadrilaterals 15–18
   parallelograms 25–31

transformation of 2D shapes 115–28
translations 223–8
trapezia 25–31
see also triangles
two-way tables 48

V

variables 41
Venn diagrams 48–9
vertically opposite angles 167

Z

the zero index 13–14